EDWARD B.

CHARLES DARWIN

AND THE

THEORY OF NATURAL SELECTION

Elibron Classics
www.elibron.com

Elibron Classics series.

© 2005 Adamant Media Corporation.

ISBN 1-4021-8493-X (paperback)
ISBN 1-4021-0985-7 (hardcover)

This Elibron Classics Replica Edition is an unabridged facsimile
of the edition published in 1896 by Cassell and Company, Ltd.,
London, Paris, and Melbourne.

Elibron and Elibron Classics are trademarks of
Adamant Media Corporation. All rights reserved.

This book is an accurate reproduction of the original. Any marks, names, colophons, imprints, logos or other symbols or identifiers that appear on or in this book, except for those of Adamant Media Corporation and BookSurge, LLC, are used only for historical reference and accuracy and are not meant to designate origin or imply any sponsorship by or license from any third party.

THE CENTURY SCIENCE SERIES

EDITED BY SIR HENRY E. ROSCOE, D.C.L., LL.D., F.R.S.

CHARLES DARWIN

AND THE

THEORY OF NATURAL SELECTION

The Century Science Series.

EDITED BY

SIR HENRY E. ROSCOE, D.C.L., F.R.S.

CASSELL & COMPANY, LIMITED, *London; Paris & Melbourne.*

Photo by Mr. James C. Christie, F.G.S., Glasgow.

STATUE OF CHARLES DARWIN.

(*By Boehm.*)

CENTRAL HALL OF THE BRITISH MUSEUM OF NATURAL HISTORY.

Charles Darwin

AND THE

THEORY OF NATURAL SELECTION

BY

EDWARD B. POULTON

M.A., F.R.S., F.G.S., F.L.S., ETC.

HOPE PROFESSOR OF ZOOLOGY AT THE UNIVERSITY OF OXFORD
CORRESP. MEMB. OF THE NEW YORK ACADEMY OF SCIENCES
CORRESP. MEMB. OF THE BOSTON SOCIETY OF NAT. HIST.

CASSELL AND COMPANY, LIMITED

LONDON, PARIS & MELBOURNE

1896

ALL RIGHTS RESERVED

INTRODUCTION

In the following pages I have tried to express a sense of the greatness of my subject by simplicity and directness of statement. The limits of the work necessarily prevented any detailed treatment, the subject of the work prevented originality. We have had the great "Life and Letters" with us for nine years, and this I have used as a mine, extracting what I believed to be the statements of chief importance for the work in hand, and grouping them so as to present what I hope is a connected account of Darwin's life, when considered in relation to his marvellous work, and especially to the great central discovery of Natural Selection and its exposition in the "Origin of Species."

In addition to the invaluable volumes which we owe to the industry, taste, and skill of Francis Darwin, an immense number of other works have been consulted. We live in an age of writing, and of speeches and addresses; and the many sides of Darwin's life and work have again and again inspired the ablest men of our time to write and speak their best—a justification for the freedom with which quotations are spread over the following pages.

It is my pleasant duty to express my hearty thanks to many kind friends who have helped in the production of this little work. Mr. Francis Darwin has kindly permitted the use of many of Darwin's letters, which have not as yet been published, and he has given me valuable information and criticism on many points. I have also gained much by discussion and correspondence with my friends Dr. A. R. Wallace, Professor E. Ray Lankester, and Professor Meldola. The latter has freely given me the use of his valuable series of letters; and I owe to my friend, Mr. Rowland H. Wedgwood, the opportunity of publishing a single letter of very great interest.

The greater part of the volume formed the subject of two short courses of lectures delivered in the Hope Department of the Oxford University Museum in Michaelmas Term 1894 and Lent Term 1895.

EDWARD B. POULTON.

Oxford, *October*, 1896.

CONTENTS

CHARLES DARWIN

AND THE

THEORY OF NATURAL SELECTION.

————◂◇▸————

CHAPTER I.

THE SECRET OF DARWIN'S GREATNESS.

CHARLES ROBERT DARWIN was born at Shrewsbury on February 12th, 1809, the year which witnessed the birth of Alfred Tennyson, W. E. Gladstone, and Abraham Lincoln.

Oliver Wendell Holmes, born in the same year, delighted to speak of the good company in which he came into the world. On January 27th, 1894, I had the great pleasure of sitting next to him at a dinner of the Saturday Club in Boston, and he then spoke of the subject with the same enthusiasm with which he deals with it in his writings; mentioning the four distinguished names, and giving a brief epigrammatic description of each with characteristic felicity. Dr. Holmes further said that he remembered with much satisfaction an occasion on which he was able to correct Darwin on a matter of scientific fact. He could not remember the details, but we may hope for their ultimate recovery, for he said that Darwin had written a courteous reply accepting the correction.

Charles Darwin's grandfather, Erasmus Darwin (1731–1802), was a man of great genius. He speculated upon the origin of species, and arrived at views which were afterwards independently enunciated by Lamarck. He resembled this great zoologist in fertility of imagination, and also in the boldness with which he put forward suggestions, many of which were crude and entirely untested by an appeal to facts. The poetical form in which a part of his work was written was, doubtless, largely due to the traditions and customs of the age in which he lived.

Robert Waring (1766–1848), the father of Charles Darwin, was the second son of Erasmus. He married a daughter of the great Josiah Wedgwood. Although his mother died when he was only eight years old, and Darwin remembered very little of her, there is evidence that she directed his attention to Nature ("Autobiography," p. 28, footnote). Dr. Darwin followed his father's profession, commencing a very successful medical practice at Shrewsbury before he was twenty-one. He was a man of great penetration, especially in the discernment of character—a power which was of the utmost value to him in his profession. Dr. Darwin had two sons and four daughters: Charles was the younger son and fourth child, his brother Erasmus being the third.

Even in this mere outline there is evidence of

hereditary genius in the Darwin family—evidence which becomes irresistible when all available details of every member of the family are brought together, as they are in the great "Life and Letters." When it is further remembered that two of Charles Darwin's sons have achieved distinction as scientific investigators, it will be admitted that the history of the family affords a most striking example of hereditary intellectual power.

There is nothing in this history to warrant the belief that the nature and direction of hereditary genius receive any bias from the line of intellectual effort pursued by a parent. We recognise the strongest evidence for hereditary capacity, but none at all for the transmission of results which follow the employment of capacity. Thus Erasmus inherited high intellectual power, with a bias entirely different from that of his younger brother Charles — his interests being literary and artistic rather than scientific. The wide difference between the brothers seems to have made a great impression upon Charles, for he wrote :—

"Our minds and tastes were, however, so different, that I do not think I owe much to him intellectually. I am inclined to agree with Francis Galton in believing that education and environment produce only a small effect on the mind of anyone, and that most of our qualities are innate" ("Life and Letters," 1887, p. 22).

Equally significant is the fact that Professor George Darwin's important researches in mathematics have been applied to astronomy — subjects which were not pursued by his father.

It appears probable that Charles Darwin's unique power was largely due to the inheritance of the imagination of his grandfather combined with the acute observation of his father. Although he possessed an even larger share of both these qualities than his predecessors, it is probable that he owed more to their co-operation than to the high degree of their development.

It is a common error to suppose that the intellectual powers which make the poet or the historian are essentially different from those which make the man of science. Powers of observation, however acute, could never make a scientific discoverer; for discovery requires the creative effort of the imagination. The scientific man does not stumble upon new facts or conclusions by accident; he finds what he looks for. The problem before him is essentially similar to that of the historian who tries to create an accurate and complete picture of an epoch out of scattered records of contemporary impressions more or less true, and none wholly true. Fertility of imagination is absolutely essential for that step from the less to the more perfectly known which we call discovery.

But fertility of imagination alone is insufficient for the highest achievement in poetry, history, or science; for in all these subjects the strictest self-criticism and the soundest judgment are necessary in order to ensure that the results are an advance in the direction of the truth. A delicately-adjusted balance between the powers of imagination and the powers which hold imagination in check, is essential in the historian who is to provide us with a picture of a past age, which explains the mistaken impression gained by a more or less prejudiced observer who saw but a small part of it from a limited standpoint, and has handed down his impression to us. A poem which sheds new light upon the relation between mind and mind, requires to be tested and controlled by constant and correct observation, like a hypothesis in the domain of the natural sciences.

It is probable, then, that the secret of Darwin's strength lay in the perfect balance between his powers of imagination and those of accurate observation, the creative efforts of the one being ever subjected to the most relentless criticism by the employment of the other. We shall never know, I have heard Professor Michael Foster say, the countless hypotheses which passed through the mind of Darwin, and which, however wild and improbable, were tested by an appeal to Nature, and were then dismissed for ever.

Darwin's estimate of his own powers is given with characteristic candour and modesty in the concluding paragraph of his " Autobiography " (" Life and Letters," 1887, p. 107) :—

" Therefore my success as a man of science, whatever this may have amounted to, has been determined, as far as I can judge, by complex and diversified mental qualities and conditions. Of these, the most important have been—the love of science—unbounded patience in long reflecting over any subject—industry in observing and collecting facts—and a fair share of invention as well as of common sense. With such moderate abilities as I possess, it is truly surprising that I should have influenced to a considerable extent the belief of scientific men on some important points."

We also know from other sources that Darwin looked upon the creative powers as essential to scientific progress. Thus he wrote to Wallace in 1857: "I am a firm believer that without speculation there is no good and original observation." He also says in the "Autobiography": "I have steadily endeavoured to keep my mind free so as to give up any hypothesis, however much beloved (and I cannot resist forming one on every subject), as soon as facts are shown to be opposed to it."

I have thought it worth while to insist thus strongly on the high value attached by Darwin to hypothesis, controlled by observation, in view of certain recent attacks upon this necessary weapon for scientific advance. Thus Bateson, in his "Materials for the Study of Variation" (London, 1894), p. 7, says: "In the old time the facts of Nature were beautiful in themselves and needed not the rouge of speculation to quicken their charm, but that was long ago before Modern Science was born." The author does not specify the period in the history of science when discovery proceeded without hypothesis. A study of the earlier volumes of the *Philosophical Transactions* reveals a far greater interest in speculation than in the facts of Nature. We can hardly call those ages anything but speculative which received with approval the suggestions that geese were developed from barnacles which grew upon trees; that swallows hibernated at the bottom of lakes; that the Trade-winds were due to the breath of a sea-weed. Bateson's statement requires to be reversed in order to become correct. Modern science differs from the science of long ago in its greater attention to the facts of Nature and its more rigid control over the tendency to hypothesis; although hypothesis remains, and must ever remain, as the guide and inspirer of observation and the discovery of fact.* Although Darwin has kindled the imagination of hundreds of workers, and has thus been the cause of an immense amount of speculation, science owes him an even larger debt for the innumerable facts discovered under the guidance of this faculty.

* See Professor Meldola's interesting Presidential Address to the Entomological Society of London (January, 1896) on the use of the imagination in science, printed in the Transactions of the Society and in *Nature*. See also "The Advancement of Science" (London, 1890), in which Professor Lankester maintains (p. 4): "All true science deals with speculation and hypothesis, and acknowledges as its most valued servant—its indispensable ally and helpmeet—that which our German friends call ' Phantasie' and we 'the Imagination.'" Consult also Professor Tyndall's essay "On the Scientific Use of the Imagination" ("Fragments of Science," 1889, vol. ii., p. 101).

CHAPTER II.

OF Darwin's boyhood and school-life we only know
the facts given in his brief "Autobiography," written
when he was sixty-seven, together with those collected
by his son Francis and appended in the form of notes.
He first went to Mr. Case's day-school in Shrewsbury
in 1817, the year of his mother's death. At this
time, although only eight years old, his interest in
natural history and in collecting was well established.
"The passion for collecting, which leads a man to be
a systematic naturalist, a virtuoso, or a miser, was
very strong in me, and was clearly innate, as none of
my sisters or brother had this taste."

In the following year he went to Dr. Butler's
school in Shrewsbury, where he remained seven years.
He does not appear to have profited much by the
classical instruction which at that time received
almost exclusive attention. His interest seems to
have been chiefly concentrated upon sport; but
whenever a subject attracted him he worked hard
at it, and it is probable that he would have conveyed
a very different impression of his powers to the
masters and his father if scientific subjects had been
taught, as they are now to a moderate extent in
many schools.

That he was a keen observer for his age is clear from the fact that, when he was only ten, he was much interested and surprised to notice that the insects he found on the Welsh coast were different from those in Shropshire. His most valuable education was received out of school hours—collecting, and working at chemistry with his brother Erasmus, although this latter study drew down upon him the rebukes of Dr. Butler for wasting time on such useless subjects.

He was removed from school early, and in 1825 went to Edinburgh to study medicine—a subject for which he seemed to be unfitted by nature. The methods of instruction by lectures did not benefit him ; he was disgusted at dissection, and could not endure to witness an operation. And yet here it was evident, as it became afterwards at Cambridge, that Darwin—although seeming to be by no means above the average when judged by ordinary standards—possessed in reality a very remarkable and attractive personality. There can be no other explanation of the impression he made upon distinguished men who were much older than himself, and the friendships he formed with those of his own age who were afterwards to become eminent.

Thus at Edinburgh he was well acquainted with Dr. Grant and Mr. Macgillivray, the curator of the museum, and worked at marine zoology in company with the former. Here, too, in 1826, he made his first scientific discovery, and read a paper before the

B

Plinian Society, proving that so-called eggs of Flustra were in reality free-swimming larvæ. And it is evident from his "Autobiography" that he took every opportunity of hearing and learning about scientific subjects.

Darwin's love of sport remained as keen as ever at this period and at Cambridge, and he speaks with especial enthusiasm of his visits in the autumn to Maer, the home of his uncle, Josiah Wedgwood, who afterwards exerted so important an influence upon his life.

After Darwin had been at Edinburgh for two sessions, his father realised that he did not like the thought of the medical profession, and suggested that he should become a clergyman. With this intention he was sent to Cambridge in the beginning of 1828, after spending some months in recovering the classics he had learnt at school.

He joined Christ's College, and passed his final examination in January, 1831, being tenth in the list of those who do not seek honours. The immense, and in many respects disastrous, development of the competitive examination system since that time has almost banished from our universities the type of student represented by Darwin—the man who takes the easiest road to a degree and obtains it with the minimum of effort, but who all the time is being benefited by residence, studying, without any thought of examinations, the subjects which are of special interest to him, and seeking personal contact with older men who have reached the highest eminence in those subjects.

He seems to have led a somewhat double life at Cambridge, his intense love of sport taking him into a pleasure-loving set, while his intellectual interests made him the intimate friend of Whitley, who became Senior Wrangler, and of Professor Henslow, to whom he was introduced by his second cousin, W. Darwin Fox, who also first interested him in entomology. He became so keen a collector of beetles that his successes and experiences in this direction seem to have impressed him more deeply than anything else at Cambridge. Entomology, and especially beetles, form the chief subject of those of his Cambridge letters which have been recovered.

Darwin's friendship with Henslow, which was to have a most important effect on his life, very soon deepened. They often went long walks together, so that he was called " the man who walks with Henslow." This fact and the subsequent rapidly formed intimacy with Professor Adam Sedgwick, indicate that he was remarkable among the young men of his standing.

One of his undergraduate friends, J. M. Herbert, afterwards County Court Judge for South Wales, retained the most vivid recollection of Darwin at Cambridge, and contributed the following impression of his character to the " Life and Letters ":—

"It would be idle for me to speak of his vast intellectual powers . . . but I cannot end this cursory and rambling sketch without testifying, and I doubt not all his surviving college friends would concur with me, that he was the most genial, warm-hearted, generous and affectionate of friends; that his sympathies were with all that was good and true; and that he had a cordial hatred for everything false, or vile, or cruel, or mean, or dishonourable. He was not only great, but pre-eminently good, and just, and loveable."

Two books greatly influenced Darwin—Herschel's "Introduction to the Study of Natural Philosophy," which, he said, "stirred up in me a burning zeal to add even the most humble contribution to the noble structure of Natural Science"; and Humboldt's "Personal Narrative," which roused in him the longing to travel—a desire which was soon afterwards gratified by his voyage in the *Beagle*.

"Upon the whole," he says, "the three years which I spent at Cambridge were the most joyful in my happy life; for I was then in excellent health, and almost always in high spirits."

After passing his last examination, Darwin had still two terms' residence to keep, and was advised by Henslow to study geology. To this end Henslow asked Sedgwick to allow Darwin to go with him on a geological excursion in North Wales in August, 1831. He thus gained experience which was of the utmost value during the voyage of the *Beagle*.

CHAPTER III.

VOYAGE OF THE "BEAGLE" (1831–36).

ABOUT the time of the excursion with Sedgwick (the exact date is uncertain) Professor Henslow received a letter from George Peacock (formerly Dean of Ely and Lowndean Professor of Astronomy at Cambridge) stating that he had the offer to recommend a young man as naturalist to accompany Captain Fitzroy on a surveying expedition to many parts of the world. Leonard Jenyns (afterwards Blomefield) was evidently considered to be the most suitable person for the position, but he was unable to accept it. Henslow at once wrote (August 24th, 1831) to Darwin, and advised him to do his utmost to obtain the position, and Darwin found the letter waiting for him on his return home after the geological excursion with Sedgwick. As his father greatly disliked the idea, Darwin at once wrote (August 30th) and declined, and the next day went to Maer to be ready for the shooting on September 1st. Here, however, his uncle, Josiah Wedgwood, took a very different view from that adopted by his father, with the result that both he and Darwin wrote (August 31st) to Shrewsbury and reopened the question. Darwin's letter shows the most touching deference to his father's wishes, and the gravest apprehension lest he should

be rendered " uncomfortable " or " uneasy " by any
further suggestion as to the possibility of the voyage,
although his father had said, " If you can find any
man of common-sense who advises you to go, I will
give my consent." We also learn from the " Autobiography " that his uncle sent for him whilst out
shooting and drove him the thirty miles to Shrewsbury, in order that they might talk with his father,
who then at once consented. This must have been
on September 1st, 1831.

From this time until he went to Plymouth, on
October 24th—the final start was not until December
27th—his letters show that he had a very busy time
making purchases and preparing for the voyage.
These letters breathe the warmest affection to the
members of his family and his friends, together with
the keenest enthusiasm for Captain Fitzroy, the ship,
and the voyage.

The voyage of the *Beagle* lasted from December
27th, 1831, to October 2nd, 1836. Darwin says that
it was " by far the most important event in my life,
and has determined my whole career. . . . I have
always felt that I owe to the voyage the first real
training or education of my mind " (*l. c.*, p. 61). He
attributes the greatest share in this training to
geology, among the special sciences, because of the
reasoning involved in making out the structure of
a new and unknown district; but he considers that
the habits of " energetic industry and of concentrated
attention " which he then acquired were of the utmost

importance, and the secret of all his success in science. He tells us that the love of sport was present at first in all its keenness, but that he gradually abandoned it for scientific work.

Among his numerous observations and discoveries during the voyage, those which appear to stand out in his mind so that he quotes them in his "Auto-biography" are—the explanation of the forms of coral islands, the geological structure of St. Helena and other islands, and the relations between the animals and plants of the several Galapagos islands to each other and to those of South America. His letters and the collections which he sent home attracted much attention; and Sedgwick told Dr. Darwin that his son would take a place among the leading scientific men. When Darwin heard this from his sisters, he says, " I clambered over the mountains of Ascension with a bounding step, and made the volcanic rocks resound under my geological hammer." His letters during the voyage are full of enthusiasm and of longing to return to his family and friends.

There was the same conflict between the naval and scientific departments of the *Beagle* on the un-tidiness of the decks which was afterwards repeated on the *Challenger*, where I have been told that one of the naval authorities used to say, with resigned disgust, " Oh, no, we're not a man-of-war, we're only a —— dredger !"

In the course of the voyage the following countries and islands were visited in the order given :—The

Cape de Verde Islands, St. Paul's Rocks, Fernando Noronha, South America (including the Galapagos Archipelago, the Falkland Islands, and Tierra del Fuego), Tahiti, New Zealand, Australia, Tasmania, Keeling Island, Maldive Coral Atolls, Mauritius, St. Helena, Ascension. Brazil was then visited again for a short time, the *Beagle* touching at the Cape de Verde Islands and the Azores on the voyage home.

Darwin says, concerning the intellectual effect of his work during the voyage :—

"That my mind became developed through my pursuits during the voyage is rendered probable by a remark made by my father, who was the most acute observer whom I ever saw, of a sceptical disposition, and far from being a believer in phrenology ; for on first seeing me after the voyage he turned round to my sisters, and exclaimed, 'Why the shape of his head is quite altered !'" (*l. c.*, pp. 63, 64).

CHAPTER IV.

DARWIN reached England October 2nd, 1836, and
was home at Shrewsbury October 5th (according to
his Letters ; the 4th is the date given by Francis
Darwin in the "Life and Letters"). The two years
and three months which followed he describes as
the most active ones he ever spent. After visiting
his family, he stayed three months in Cambridge,
working at his collection of rocks, writing his
"Naturalist's Voyage," and one or two scientific
papers. He then (March 7th, 1837) took lodgings
in 36, Great Marlborough Street, London, where he
remained until his marriage, January 29th, 1839.
The apathy of scientific men—even those in charge
of museums—caused him much depression, and he
found great difficulty in getting specialists to work
out his collections, although the botanists seem to
have been keener than the zoologists.

The commencement of his London residence is of
the deepest interest, as the time at which he began
to reflect seriously on the origin of species. Thus he

says in the "Autobiography":—"In July I opened my first note-book for facts in relation to the Origin of Species, about which I had long reflected, and never ceased working for the next twenty years." Furthermore, his pocket-book for 1837 contained the words:—"In July opened first note-book on Transmutation of Species. Had been greatly struck from about the month of previous March" (he was then just over twenty-eight years old) "on character of South American fossils, and species on Galapagos Archipelago. These facts (especially latter) origin of all my views." It is, perhaps, worth while to explain in greater detail the nature of this evidence which appealed so strongly to Darwin's mind. The Edentata (sloths, ant-eaters, armadilloes, etc.) have their metropolis in South America, and in the later geological formations of this country the skeletons of gigantic extinct animals of the same order (Megatherium, Mylodon, Glyptodon, etc.) are found; and Darwin was doubtless all the more impressed by discovering such remains for himself. In his "Autobiography" he says: "During the voyage of the *Beagle* I had been deeply impressed by discovering in the Pampean formation great fossil animals covered with armour like that on existing armadilloes; . . . ."

Darwin was thus led to conclude that there was some genetic connection between the animals which have succeeded each other in the same district; for in a theory of destructive cataclysms, followed by re-creations—or, indeed, in any theory of special creation—there seemed no adequate reason why the successive forms should belong to the same order. In his "Naturalist's Voyage Round the World" he says, speaking of this subject: "This wonderful relationship in the same continent between the dead and the living will, I do not doubt, hereafter throw more light on the appearance of organic beings on our earth, and their disappearance from it, than any other class of facts" (p. 173 in the third edition).

The other class of evidence which impressed him even more strongly was afforded by the relations between the animals and plants of the several islands of the Galapagos Archipelago and between those of the Archipelago and of South America, nearly 600 miles to the East. Although the inhabitants of the separate islands show an astonishing amount of peculiarity, the species are nearly related, and also exhibit American affinities. Concerning this, Darwin writes in his "Voyage" (p. 398 in the third edition): "Reviewing the facts here given, one is astonished at the amount of creative force—if such an expression may be used—displayed on these small, barren, and rocky islands; and still more so at its diverse and yet analogous action on points so near each other." Here, too, the facts were unintelligible on a theory of separate creation of species, but were at once explained if we suppose that the inhabitants were the modified descendants of species which had migrated from South America — the migrations to the Archipelago and between the separate islands being rendered extremely rare from the depth of the sea, the direction of the currents, and the absence of gales. In this way time for specific modification was provided before the partially modified form could interbreed with the parent species and thus lose its own newly-acquired characteristics.

Although Darwin made these observations on the *Beagle*, they required, as Huxley has suggested (Obituary [1888], "Darwiniana": Collected Essays, vol. ii., pp. 274–275. London, 1893), careful and systematic working out before they could be trusted as a basis on which to speculate ; and this could not be done until the return home. The following letter written by Darwin to Dr. Otto Zacharias in 1877 confirms this opinion. It was sent to Huxley by Francis Darwin, and is printed in "Darwiniana" (*l. c.*, p. 275) :—

"When I was on board the 'Beagle,' I believed in the permanence of species, but, as far as I can remember, vague doubts occasionally flitted across my mind. On my return home in the autumn of 1836, I immediately began to prepare my journal for publication, and then saw how many facts indicated the common descent of species, so that in July, 1837, I opened a note-book to record any facts which might bear on the question. But I did not become convinced that species were mutable until I think two or three years had elapsed."

It is interesting to note that both the lines of evidence which appealed to Darwin so strongly, point to evolution, but not to any causes of evolution. The majority of mankind were only convinced of this process when some conception as to its causes had been offered to them ; Darwin took the more logical course of first requiring evidence that the process takes place, and then inquiring for its causes.

The first indication of these thoughts in any of his published letters is in one to his cousin Fox written in June, 1838, in which, after alluding to some questions he had previously asked about the crossing of animals, he says, " It is my prime hobby, and I really think some day I shall be able to do something in that most intricate subject—species and varieties."

He is rather more definite in a letter to Sir Charles Lyell, written September 13th in the same year :—

" I have lately been sadly tempted to be idle—that is, as far as pure geology is concerned—by the delightful number of new views which have been coming in thickly and steadily,—on the classification and affinities and instincts of animals—bearing on the question of species. Note-book after note-book has been filled with facts which begin to group themselves *clearly* under sub-laws."

On February 16th, 1838, he was appointed Secretary of the Geological Society, a position which he retained until February 1st, 1841. During these two years after the voyage he saw much of Sir Charles Lyell, whose teachings had been of the greatest help to him during the voyage, and whose method of appealing to natural causes rather than supernatural cataclysms undoubtedly had a most

important influence on the development of Darwin's
mind.   This · influence he delighted to acknowledge, dedicating to Lyell the second edition of his
" Voyage," " as an acknowledgment that the chief
part of whatever scientific merit this ' Journal '
and the other works of the author may possess has
been derived from studying the well-known and
admirable ' Principles of Geology.' "

At this period he finished his " Journal," which
was published in 1839 as Vol. III. of the " Narrative
of the Surveying Voyages of Her Majesty's Ships
Adventure and *Beagle*."   A second edition was published in a separate form in 1845 as the " Journal of
Researches into the Natural History and Geology of
the Countries visited during the Voyage of H.M.S.
Beagle round the World, under the command of
Captain Fitz-Roy, R.N."; and a third édition—but
very slightly altered—in 1860, under the title " A
Naturalist's Voyage : Journal of Researches, etc."
This book is generally admitted to deserve above all
others the generous description which Darwin gave
to Sir Joseph Hooker of Belt's admirable " Naturalist
in Nicaragua "—as " the best of all Natural History
journals which have ever been published."

A comparison between the first and second
editions indicates, but by no means expresses, his
growing convictions on evolution and natural selection.   Natural selection he had not discovered when
the MS. of the first edition was complete; and if we
had no further evidence we could not, from any

passage in the work, maintain that he was convinced of evolution. His great caution in dealing with so tremendous a problem explains why the second edition does not reflect the state of his mind at the time of its publication. He tells us (" Autobiography ") that in the preparation of this second edition he "took much pains," and we may feel confident that much of this care was given to the decision as to how much he should reveal and how much withhold of the thoughts which were occupying his mind, and the conclusions to which he had at that time arrived. That he did attribute much importance to the evolutionary passages added in the second edition is shown by his letter to Lyell (July, 1845), in which he alludes to some of them, and specially asks Lyell to read the pages on the causes of extinction.

He also edited and superintended the " Zoology of the Voyage of H.M.S. *Beagle*," the special parts of which were written by various eminent systematists, and appeared separately between 1839 and 1843.

He also read several papers before the Geological Society, including two (1838 and 1840) on the Formation of Mould by the Action of Earth-Worms—a subject to which he returned, and upon which his last volume (published in 1881) was written. He also read a paper on the Parallel Roads of Glen Roy before the Royal Society (published in the *Phil. Trans.*, 1839). These wonderful parallel terraces are now admitted to be due to the changes of level in a lake following those of an ice-barrier at the mouth of the valley. At the time Darwin studied them, the terraces were believed to have been formed by a lake dammed back by a barrier of rock and alluvium; this he proved to be wrong, and as no other barrier was then available—for the evidences of glaciation had not then been discovered by Agassiz—he was driven, on the method of exclusion, to the action of the sea. Upon this subject he says, in the " Autobiography," " My error has been a good lesson to me never to trust in science to the principle of exclusion."

On January 29th, 1839, he married his cousin, Emma Wedgwood, the daughter of Josiah Wedgwood, of Maer. They resided at 12, Upper Gower Street until September 14th, 1842, when they settled at Down.

The few graceful and touching words in which Francis Darwin, in the " Life and Letters," alludes to his father's married life show how deep is the debt of gratitude which the world owes to Mrs. Darwin; for without her constant and loving care it would have been impossible for Darwin to have accomplished his life-work.

During these years in London his health broke down many times; so that he says, in the " Autobiography": " I did less scientific work, though I worked as hard as I possibly could, than during any other equal length of time in my life." He chiefly worked at his book on " The Structure and Distribution of Coral Reefs," published in 1842 (second edition in 1874). This work contains an account of Darwin's well-known theory upon the origin of the various coral formations—fringing reefs, barrier reefs, and atolls—by the upward growth of the reef keeping pace with the gradual sinking of the island upon which it is based, so that the living corals always remain at the surface under the most favourable conditions, while beneath them is an ever-thickening reef formed of dead coral, until at length, by continuing this process, the climax is reached in the atoll, in which the original island has altogether disappeared beneath the surface of a central lagoon enclosed in a ring formed by the living edge of the reef. This theory, after being accepted for many years, has recently been disputed, chiefly as the result of the observations made on the *Challenger* expedition. It is contended by Dr. John Murray " that it is not necessary to call in subsidence to explain any of the characteristic features of barrier reefs or atolls, and that all these features would exist alike in areas of slow elevation, of rest, or of slow subsidence " (*Nature*, August 12th, 1880, p. 337). It cannot be said that this controversy is yet settled, or that the supporters of either theory have proved that the other does not hold—at any rate, in certain cases.

Among his geological papers written at this time was one describing the glacial phenomena observed during a tour in North Wales. This paper (*Philosophical Magazine*, 1842, p. 352) is placed by Sir

c

Archibald Geikie "almost at the top of the long list
of English contributions to the history of the Ice
Age."

At this time, too, he was reflecting and collecting
evidence for the great work of his life.   Thus in
January, 1841, he writes to his cousin, Darwin Fox,
asking for "all kinds of facts about 'Varieties and
Species.'"

CHAPTER V.

FROM September 14th, 1842, until his death, Darwin resided at Down, living a very retired life, and almost exclusively engaged in his scientific researches. Although Down is only twenty miles from London, it is three miles from the nearest railway station (Orpington), and is only now for the first time receiving a telegraph office. A home in such a place enabled Darwin to pursue his work without interruption, remaining, at the same time, within easy reach of all the advantages of London. Here, too, he had no difficulty in avoiding social engagements, which always injured his very precarious health, and thus interfered with work; although, at the same time, he could entertain in his own house at such times as he felt able to do so.

In 1844, and again in 1846, he published works on the geology of the voyage of the *Beagle;* the first on the Volcanic Islands visited, the second on South America. A second edition, in which both were combined in a single work, appeared in 1876. He seemed somewhat disappointed at the small amount of attention they at first attracted, and wrote with much

humour to J. M. Herbert:—" I have long discovered
that geologists never read each other's works, and
that the only object in writing a book is a proof of
earnestness, and that you do not form your opinions
without undergoing labour of some kind." All geologists were, nevertheless, soon agreed in attaching the
highest value to these researches.

From this time forward his work was almost
exclusively zoological. The four monographs on the
Cirripedia, recent and fossil, occupied eight years—
from October, 1846, to October, 1854. The works on
the recent forms were published by the Ray Society
(1851 and 1854), and those on the fossil forms by the
Palæontographical Society (1851 and 1854). These
researches grew directly out of his observations on
the *Beagle*, but it is evident that they reached far
greater dimensions than he had at first intended.
Thus, at the very beginning of the work, he wrote
(October, 1846) to Hooker:—

"I am going to begin some papers on the lower marine
animals, which will last me some months, perhaps a year, and
then I shall begin looking over my ten-year-long accumulation
of notes on species and varieties, which, with writing, I
dare say will take me five years, and then, when published, I
dare say I shall stand infinitely low in the opinion of all sound
Naturalists—so this is my prospect for the future."

Darwin himself, at any rate towards the end of
his life, when he wrote his " Autobiography," doubted
" whether this work was worth the consumption of
so much time," although admitting that it was of
" considerable value" when he had "to discuss in

the 'Origin of Species' the principles of a natural classification." Sir Joseph Hooker remembers that Darwin at an earlier time "recognised three stages in his career as a biologist: the mere collector at Cambridge; the collector and observer in the *Beagle* and for some years afterwards; and the trained naturalist after, and only after, the Cirripede work" (Letter to F. Darwin).

Professor Huxley considers that just as by Darwin's practical experience of physical geography, geology, etc., on the *Beagle*, "he knew of his own knowledge the way in which the raw materials of these branches of science are acquired, and was, therefore, a most competent judge of the speculative strain they would bear," so his Cirripede work fitted him for his subsequent speculations upon the deepest biological problems. "It was a piece of critical self-discipline, the effect of which manifested itself in everything your father wrote afterwards, and saved him from endless errors of detail" (Letter to F. Darwin, "Life and Letters"). The history of Darwin's career has often been used as an argument against those who, not having passed through a similar training as regards systematic zoological work, have ventured to concern themselves with the problems of evolution. Professor Meldola has recently treated of this subject in his interesting presidential address to the Entomological Society (1896). He says:—

"It used formerly to be asserted that he only is worthy of attention who has done systematic, *i.e.* taxonomic, work. I do not know whether this view is still entertained by entomologists; if so, I feel bound to express my dissent. It has been pointed out that the great theorisers have all done such work —that Darwin monographed the Cirripedia, and Huxley the oceanic Hydrozoa, and it has been said that Wallace's and Bates's contributions in this field have been their biological salvation. I yield to nobody in my recognition of the value and importance of taxonomic work, but the possibilities of biological investigation have developed to such an extent since Darwin's time that I do not think this position can any longer be seriously maintained. It must be borne in mind that the illustrious author of the 'Origin of Species' had none of the opportunities for systematic training in biology which any student can now avail himself of. To him the monographing of the Cirripedia was, as Huxley states in a communication to Francis Darwin, 'a piece of critical self-discipline,' and there can be no reasonable doubt that this value of systematic work will be generally conceded. That this kind of work gives the sole right to speculate at the present time is, however, quite another point."

Meldola then goes on to argue that the systematic work of those who know nothing of the living state of the species they are describing does not specially fit them for theorising, and he concludes by quoting the following passage from a letter recently received from A. R. Wallace :—

"I do not think species-describing is of any special use to the philosophical generaliser, but I do think the collecting, naming, and classifying some extensive group of organisms is of great use, is, in fact, almost essential to any thorough grasp of the whole subject of the evolution of species through variation and natural selection. I had described nothing when I wrote my papers on variation, etc. (except a few fishes and palms from the Amazon), but I had collected and made out species very largely and had seen to some extent how curiously useful and protective their forms and colours often were, and all this was of great use to me."

Towards the end of this long period of hard taxonomic labour, we know from Darwin's letters that he was extremely tired of the work; but with marvellous resolution — and in spite of the trouble of his health, which was perhaps worse than at any other time—he clung to and carried through this stupendous task, although all the time attracted away from it by the weightier problems which he could never thrust aside after they had once made their claim upon him.

Darwin was evidently greatly disconcerted at the task of making out those special difficulties which man has added to the difficulties of Nature herself— the disheartening tangle of nomenclature. He thought that the custom of appending the name of the systematist after that of the species or genus he had named was injurious to the interests of science— inducing men to name quickly rather than describe accurately. Some of his remarks on this subject indicate the state of his mind. Thus he wrote to Hooker, October 6th, 1848 :—

" I have lately been trying to get up an agitation . . . against the practice of Naturalists appending for perpetuity the name of the *first* describer to species. I look at this as a direct premium to hasty work, to *naming* instead of *describing*. A species ought to have a name so well known that the addition of the author's name would be superfluous, and . . empty vanity. . . . Botany, I fancy, has not suffered so much as zoology from mere *naming;* the characters, fortunately, are more obscure. . . . Why should Naturalists append their own names to new species, when Mineralogists and chemists do not do so to new substances ? "

And again he wrote to Hugh Strickland, January 29th, 1849 :—

"I have come to a fixed opinion that the plan of the first describer's name, being appended for perpetuity to a species, has been the greatest curse to Natural History. . . . I feel sure as long as species-mongers have their vanity tickled by seeing their own names appended to a species, because they miserably described it in two or three lines, we shall have the same *vast* amount of bad work as at present, and which is enough to dishearten any man who is willing to work out any branch with care and time."

And in another letter (February 4th) to the same correspondent :—

"In mineralogy I have myself found there is no rage to merely name ; a person does not take up the subject without he intends to work it out, as he knows that his *only* claim to merit rests on his work being ably done, and has no relation whatever to *naming*. . . . I do not think more credit is due to a man for defining a species, than to a carpenter for making a box. But I am foolish and rabid against species-mongers, or rather against their vanity ; it is useful and necessary work which must be done ; but they act as if they had actually made the species, and it was their own property."

A little later in the same year (1849) his health seems to have determined him to give up the crusade, for he writes to Hooker (April 29th):—

"With health and vigour, I would not have shewn a white feather, [and] with aid of half-a-dozen really good Naturalists, I believe something might have been done against the miserable and degrading passion of mere species naming."

Anyone whose researches have been among the species of any much-worked and much-collected zoological group will quite agree that synonymy is, as

Darwin found it, heart-breaking work; and although there may be good reasons why the system of appending the describer's name must be retained, such a protest as that raised in these letters cannot fail to do good in drawing attention to an abuse which is only too common, and which introduces unnecessary difficulty and gratuitous confusion into the study of Nature.

His father, Dr. Darwin, died November 13th, 1848, at the age of eighty-three, when he was so much out of health that he was unable to attend the funeral. In 1851 he lost his little daughter Annie, who died at Malvern, April 23rd. A few days after her death he wrote a most affecting account of her— a composition of great beauty and pathos.

CHAPTER VI.

IN dealing with this subject in his "Autobiography," Darwin tells us of his reflections whilst on the voyage of the *Beagle*, and here mentions another observation which deeply impressed him in addition to those which he again repeats, on the relation between the living and the dead in the same area and on the productions of the Galapagos Archipelago—viz. "the manner in which closely allied animals replace one another in proceeding southwards over the continent" (of South America). On the theory of separate creation the existence of such representative species received no explanation, although it became perfectly intelligible on the theory that a single species may be modified into distinct, although nearly related, species in the course of its range over a wide geographical area. Here, too, the evidence is in favour of evolution simply, and does not point to any cause of evolution.

He also implies that even at this time he regarded the beautiful adaptations or contrivances of nature by which organisms are fitted to their habits of life— "for instance, a wood-pecker or a tree-frog to climb trees, or a seed for dispersal by hooks or plumes"—as the most striking and important phenomena of the organic world, and the one great difficulty in the path of any naturalist who should attempt to supply a motive force for evolution. And he regarded the previous attempts at an explanation—the direct action of surroundings and the will of the organism—as inadequate because they could not account for such adaptations.

Therefore being convinced of evolution, but as yet unprovided with a motive cause which in any way satisfied him, he began in July, 1837, shortly after his return home from the *Beagle*, to collect all facts which bore upon the modifications which man has induced in the animals and plants which he has subjugated, following, as he tells us, the example of Lyell in geology. He goes on to say in his " Autobiography " :—

" I soon perceived that selection was the key-stone of man's success in making useful races of animals and plants. But how selection could be applied to organisms living in a state of nature remained for some time a mystery to me."

We see indications in the extracts from his note-book at this period (viz. between July, 1837, and February, 1838), and before he had arrived at the conception of Natural Selection, that he had the idea of " laws of change " affecting species to some extent like the laws of change which compel the individuals of every species to work out their own development, the extinction of the one corresponding in a measure to the death of the other. Thus he says, " It is a wonderful fact, horse, elephant, and mastodon dying out about the same time in such different quarters. Will Mr. Lyell say that some [same ?] circumstance killed it over a tract from Spain to South America ?

Never." We know that a few months later he would have himself accepted the view he imputes to Lyell, and would have regarded the extinction as due to some circumstance affecting the competition for food or some other relationship with the organic life of the same district. It is probable that the above quotation from his Diary was written in connection with the conclusion of Chapter IX. of the first edition of the "Journal of the Voyage" (pp. 211, 212); for the latter is a fuller exposition of the same argument.*

" One is tempted to believe," he says, "in such simple relations, as variation of climate and food, or introduction of enemies, or the increased numbers of other species, as the cause of the succession of races. But it may be asked whether it is probable that ["than" is an evident misprint in the original] any such cause should have been in action during the same epoch over the whole northern hemisphere, so as to destroy the *Elephas primigenius* on the shores of Spain, on the plains of Siberia, and in Northern America. . . . These cases of extinction forcibly recall the idea (I do not wish to draw any close analogy) of certain fruit-trees, which, it has been asserted, though grafted on young stems, planted in varied situations, and fertilized by the richest manures, yet at one period have all withered away and perished. A fixed and determined length of life has in such cases been given to thousands and thousands of buds (or individual germs), although produced in long succession."

He then concludes that the animals of one species, although " each individual appears nearly independent of its kind," may be bound together by common laws.

* " We are told in the " Life and Letters " that the last proof of the " Journal " was finished in 1837. The Diary, as stated above, was written between July, 1837, and February, 1838.

He ends by arguing that the adaptations of animals confined to certain areas cannot be related to the peculiarities of climate or country, because other animals introduced by man are often so much more successful than the aborigines. As to the causes of extinction, "all that at present can be said with certainty is that, as with the individual, so with the species, the hour of life has run its course, and is spent."

At this time he had the conception—as we see in the succeeding extracts from his Diary—of species being so constituted that they must give rise to other species; or, if not, that they must die out, just as an individual dies unrepresented if it has no offspring; that change—and evidently change in some fixed direction—or extinction, is inevitable in the history of a species after a certain period of time. With this view, which presented much resemblance to that of the author of the "Vestiges," and which seemed uppermost in his mind at this time, there are traces of others. Thus in one extract the "wish of parents" was thought of as a very doubtful explanation of adaptation, while in another we meet a tolerably clear indication of natural selection, a variety which is not well adapted being doomed to extinction, while a favourable one is perpetuated, the death of a species being regarded as "a consequence . . . of non-adaptation of circumstances."

It seems certain that for fifteen months after July, 1837, he was keenly considering the various causes of evolution which were suggested to him by the facts of nature, and that some general idea of natural selection presented itself to him at times, although without any of the force and importance it assumed in his mind at a later time.

In October, 1838, he read " Malthus on Population," and as he says:—

"Being well prepared to appreciate the struggle for existence which everywhere goes on from long-continued observation of the habits of animals and plants, it at once struck me that under these circumstances favourable variations would tend to be preserved, and unfavourable ones to be destroyed. The result of this would be the formation of new species. Here then I had a theory by which to work."

In June, 1842, he wrote a brief account of the theory, occupying thirty-five pages. In Lyell's and Hooker's introduction to the joint paper by Darwin and Wallace in the Linnean Society's Journal (1858) it is stated that the first sketch was made in 1839, but Francis Darwin shows (" Life and Letters," 1887, Vol. II. pp. 11, 12) that in all probability this is an error—a note of Darwin's referring to the first complete grasp of the theory after reading Malthus, being mistaken for a reference to the first written account.

In 1844 the sketch was enlarged to a written essay occupying 231 pages folio—" a surprisingly complete presentation of the argument afterwards familiar to us in the 'Origin of Species'" published fifteen years later. Professor Huxley, after reading this essay, observed that " much more weight is attached to the influence of external conditions in producing variation, and to the inheritance of acquired habits than in the 'Origin,'" while Professor Newton pointed out that the remarks on the migration of birds anticipate the views of later writers.*

The explanation of divergence of species during modification (divergence of character) had not then occurred to him, and he tells us in the "Autobiography :—

"I can remember the very spot in the road, whilst in my carriage, when to my joy the solution occurred to me ; and this was long after I had come to Down. The solution, as I believe, is that the modified offspring of all dominant and increasing forms tend to become adapted to many and highly diversified places in the economy of nature."

A good example of this tendency is seen in the relations of three great vertebrate classes— mammals, birds, and fishes — to the environments for which they are respectively fitted : earth, air, and water. Competition is most severe between forms most nearly alike, and hence some measure of relief from competition is afforded when certain members of each of these classes enter the domain of one of the others. Hence, we observe that although mammals as a whole are terrestrial, a small minority are aërial and aquatic ; although birds are aërial, a minority are terrestrial and aquatic ; although fishes are aquatic, a minority tend to be, at any rate largely, terrestrial and aërial.

* Professor H. F. Osborn has rightly urged that this essay should be published ("From the Greeks to Darwin," 1894, p. 235).

Huxley considered it "curious that so much importance should be attached to this supplementary idea. It seems obvious that the theory of the origin of species by natural selection necessarily involves the divergence of the forms selected" ("Obituary," 1888, reprinted in "Darwiniana," 1893; see pp. 280, 281). But Darwin showed that divergence might be a great advantage in itself, and would then be directly (and not merely incidentally and indirectly) encouraged and increased by natural selection.

As soon as the 1844 sketch was finished, Darwin wrote a letter (July 5th) as his "solemn and last request" that his wife would, in the case of his death, devote £400, or if necessary £500, in publishing it, and would take trouble in promoting it. He suggests Lyell as the best editor, then Edward Forbes, then Henslow ("quite the best in many respects"), then Hooker ("would be *very* good"), then Strickland. After Strickland he had thought of Owen as "very good," but added, "I presume he would not undertake such a work." If no editor could be obtained, he requested that the essay should be published as it was—stating that it was not intended for publication in its present form. In August, 1854, he wrote on the back of the letter: "Hooker by far best man to edit my Species volume."

All this shows how certain he felt that he was on firm ground, and that his theory of natural selection was of vast importance to science. This same strong conviction appears clearly in the first edition of the " Origin," and is undoubtedly one of the secrets of its power to move the minds of men. Although the author is above all others fair-minded; although he is most keen to discover and to bring forward all opposing evidence, and to criticise most minutely everything favourable; nevertheless, looking at the evidence as a whole, he has no doubt as to its bearing, and feels, and shows that he feels, a magnificent confidence in the truth and the importance of his theory.

CHAPTER VII.

GROWTH OF THE "ORIGIN" (*continued*) — CORRE-
SPONDENCE WITH FRIENDS.

THE great periods of Darwin's scientific career are
marked by intimate friendships, which must be
taken into account in attempting to trace his mental
development. Henslow was his intimate friend at
Cambridge and during the voyage of the *Beagle*.
The influence of Lyell, through his writings, was of
the utmost importance during the voyage, and was
deepened by the close personal contact which took
place on Darwin's return. Sir Joseph Hooker was
his most intimate friend during the growth of the
" Origin of Species."

Although Hooker met Darwin in 1839, their
friendship did not begin until four years later, when
the former returned from the Antarctic Expedition.
On January 11th, 1844, Darwin wrote admitting his
conclusions on the question of evolution:—"At last
gleams of light have come," he says, "and I am
almost convinced (quite contrary to the opinion I
started with) that species are not (it is like con-
fessing a murder) immutable" ("Life and Letters,"
Vol. II. p. 23).

From this point onwards his letters, especially to

Hooker, indicate the course he was following and the various problems he was considering as they arose. Thus we find that he had finished reading Wollaston's "Insecta Maderensia" in 1855 (writing March 7th), and had been struck with the very large proportion of wingless beetles, and had interpreted the observation, viz. "that powers of flight would be injurious to insects inhabiting a confined locality, and expose them to be blown to the sea." It is of great interest thus to witness the origin of a theory which has since been universally accepted, and has received confirmation from many parts of the world.

On April 11th of the same year he is experimenting on the powers of resistance to immersion in salt water possessed by seeds, and he writes an account of it to Hooker. The object of these experiments was to throw light on the means by which plants have been transported to islands.

In the same year began his correspondence with Asa Gray, who soon became one of his warmest friends. He had numerous questions to ask about the geographical range of plants, and in 1857 he wrote explaining in some detail the views at which he had arrived as to the causes of evolution.

My friend Rowland H. Wedgwood, a nephew of Darwin, has given me the following interesting letter to his father, which was written, he believes, probably before 1855. By kind permission, it is here published for the first time. The letter is of great interest, as throwing light upon his work, and also because of this early reference to Huxley :—

"Down, *Sept.* 5.

"MY DEAR HARRY,—I am very much obliged for the Columbine seed and for your note which made us laugh heartily.

"I had no idea what trouble the counting must have been, I had not the least conception that there would have been so many pods. I am very much interested on this point, and therefore to make assurance sure, I repeat your figures viz. 560 and 742 pods on two plants and 7200 on another. Does the latter number really mean pods and not seeds? Upon my life I am sorry to give so much trouble, but I should be VERY MUCH obliged for a few *average* size pods, put up separately that I may count the seeds in each pod: for though I counted the seeds in the pods sent before, I hardly dare trust them without counting more. Moreover I sadly want more seed itself for one of my experiments.

"The young cabbages are coming up already. Thank you much about the asparagus seeds; as it is so rare a plant, you are my only chance.

"We have been grieved to hear about poor Anne and Tom.—"Your affect's screw   "C. DARWIN.

"Have you been acquainted with Mr. Huxley; I think you would find him a pleasant acquaintance. He is a very clever man."

Mr. Francis Darwin believes that the asparagus and cabbage seeds were for the experiments to determine the time during which immersion in salt water could be endured. The object of such experiments was to throw light on the means by which plants are distributed over the earth's surface. He also informs me that the use of the word "screw" is unique and incomprehensible.

Darwin tells us in the "Autobiography" that

"early in 1856 Lyell advised me to write out my
views pretty fully, and I began at once to do so on a
scale three or four times as extensive as that which
was afterwards followed in my 'Origin of Species.'"
This work he began on May 14th, and, after working
steadily until June, 1858, had written about half the
book, in ten chapters, when he received the celebrated
letter from Wallace, which altered everything.

At this period we get interesting evidence of his
extraordinary insight in the strong protests he makes
against the Atlantis hypothesis of Edward Forbes,
and the other vast continental extensions which
naturalists did not hesitate to make in order to
explain the existence of species common to countries
separated by wide tracts of the ocean. These lost
continents were as generally accepted as they were
freely proposed. And yet we find that, even then,
one thinker far ahead of his time saw clearly enough
—as the *Challenger* Expedition twenty years later
proved beyond all doubt—that the geological evidence is against such extension, and that the means
of distribution possessed by animals are such as to
render the supposition unnecessary.

In June, 1856, he writes to Lyell: "My blood gets
hot with passion and turns cold alternately at the
geological strides, which many of your disciples are
taking"; and after mentioning the extension of
continents proposed by many leading naturalists, he
says: "If you do not stop this, if there be a lower
region for the punishment of geologists, I believe, my

great master, you will go there. Why, your disciples in a slow and creeping manner beat all the old Catastrophists who ever lived! You will live to be the great chief of the Catastrophists." Lyell wrote disagreeing on the subject of continental extension; and hence, on June 25th, 1856, Darwin replied in a long letter, giving in detail his reasons for rejecting the hypothesis. He argued (1) that the supposed extension of continents and fusion of islands would be vast changes, giving the earth a new aspect, but that recent and tertiary molluscs, etc., are distinct on opposite sides of the existing continents; so that, although he did not doubt *great* changes of level in parts of continents, he concluded that "*fundamentally* they stood as barriers to the sea where they now stand" ever since the appearance of living species; (2) that if a continent were nearly submerged, the last remaining peaks would by no means always be volcanic, as are, almost without exception, the oceanic islands; (3) that the amount of subsidence which took place in continental areas during the Silurian and Carboniferous periods—viz. during one tolerably uniform set of beings—would not be enough to account for the depth of the ocean over some parts of the site of the supposed submerged continents; (4) that the supposed extensions are not consistent with the *absence* of many groups of animals—*e.g.* mammals, frogs, etc.—from islands.

These arguments did not convince Lyell; and they have only received an almost universal acceptance after the confirmatory evidence afforded by the voyage of the *Challenger*. Dredgings over many parts of the ocean showed that all the continental deposits are collected on a fringing shelf not more than 200 miles wide, and that beyond this in the ocean bed proper an entirely different kind of deposit is accumulating, composed of the shells, bones, and teeth of swimming or floating organisms, or the products of their decomposition, of volcanic and cosmic dust, and the products—*e.g.* manganese dioxide—of the decomposition of these and of floating pumice. Hence, the depths of the ocean afford no indications of a lost continental area, but are covered by a peculiar deposit unknown among the rocks of continents which were formed in comparatively shallow water round and not far from coasts, or in land-locked or nearly land-locked seas like the Mediterranean.

On July 20th, 1856, he wrote to Asa Gray, giving some account of his views, and stating his belief in evolution, but only hinting at natural selection.

About this time we meet with evidence of the great difficulty with which Darwin's ideas were thoroughly understood, even by his intimate friends, to whom he often wrote on the subject. Later on, when the "Origin of Species" was published, although the arguments in favour of natural selection were given in considerable detail, many years passed before the theory itself was understood by the great body of naturalists. This particular case of misunderstanding is of such great interest that it is desirable to consider it in detail.

In the origin of new species by natural selection, the stress of competition determines the survival of favourable individual variations, and these, when by the continued operation of the process they have become constant, are added to those pre-existing characters of the species which are inherited from a remote past, and are witnesses of the operation of natural selection from age to age under ever-changing conditions of competition and variation. It follows, therefore, that the origin of a species can only take place once; for it is infinitely improbable that the same variation would be independently submitted under the same conditions of competition, and added to the mass of inherited characters independently gained in two distinct lines by natural selection acting in the same manner upon the same variations in the same order through all ages. Not only is it inconceivable that the same species could arise by natural selection from distinct lines of ancestry, but it is extremely improbable that the same species could arise independently in more than one centre among the individuals of a changing species; for in this case, too, it is most unlikely that the same conditions of competition would co-exist with the same favourable variations in the areas inhabited by independent colonies of the same species.

Under other theories of evolution—direct action of environment, supposed inherited effects of use and disuse, etc.—an independent origin, even from quite distinct lines, would be probable ; and we find, accordingly, that those who would advance such theories believe in what is called the "polyphyletic" origin of species (*e.g.* the horse), and in the principle of "convergence" carried far enough to produce the same complex character (*e.g.* vertebrate teeth) twice over without any genetic connection between the forms in which the character appears.

Under natural selection, however, such a result would be infinitely improbable, and hence this theory strongly supports, and indeed explains, the theory of "specific centres," viz. that each species has arisen in one area only, and has spread from that into the other areas over which it now occurs. This view was strongly held by Lyell and Hooker after an exhaustive study of the facts then known as to the geographical distribution of plants and animals ; and yet both of these distinguished naturalists seem to have feared that Darwin, in advancing a theory which was entirely consistent with their convictions and utterly inconsistent with any other views upon the same subject, was in some way undermining the conclusions at which they had arrived.

Thus Lyell wrote (July 25th, 1856) to Hooker :—

"I fear much that if Darwin argues that species are phantoms, he will also have to admit that single centres of dispersion are phantoms also, and that would deprive me of much of the value which I ascribe to the present provinces of animals and plants, as illustrating modern and tertiary changes in physical geography."

And on August 5th of the same year Darwin replied to Hooker, who had apparently argued that the origin of species by direct action of climate, etc., would mean independent and multiple specific centres :—

"I see from your remarks that you do not understand my notions (whether or no worth anything) about modification ; I attribute very little to the direct action of climate, etc. I suppose, in regard to specific centres, we are at cross purposes ; I should call the kitchen garden in which the red cabbage was produced, or the farm in which Bakewell made the Shorthorn cattle, the specific centre of these *species!* And surely this is centralisation enough !"

As I have argued above, Darwin was all the time affording the strongest support to the theory of specific centres : support which was entirely wanting in the theory of separate creation, in which the origin of each species is wrapped in mystery, so that we can form no opinion as to whether it took place at one centre or at many.

At this time, when the views set forth in the "Origin" were gaining shape and expression, we cannot estimate too highly the value of the correspondence with Hooker. In after years, when the "Origin" had to stand the fire of adverse criticism, and at first of very general disapproval, it was of inestimable advantage that every idea contained in it should have been minutely discussed beforehand with one who was more critical and more learned than the greatest of those who afterwards objected. Darwin tells us in his " Autobiography ":—

" I think that I can say with truth that in after years, though I cared in the highest degree for the approbation of such men as Lyell and Hooker, who were my friends, I did not care much about the general public."

But, although Darwin cared nothing for it, it is nevertheless true that the approbation of minds such as these was a sure indication of the general approbation of the intellect of the country, and of the world, which was to follow as soon as the new ideas were absorbed.

And the value which Darwin himself placed on these discussions appears again and again in his letters. To take a single example, he writes to Hooker November 23rd, 1856 :—

" I fear I shall weary you with letters, but do not answer this, for in truth and without flattery, I so value your letters, that after a heavy batch, as of late, I feel that I have been extravagant and have drawn too much money, and shall therefore have to stint myself on another occasion."

CHAPTER VIII.

DARWIN AND WALLACE (1858).

THE history of Darwin's friendship with Alfred
Russel Wallace is of quite unique interest, being
brought about by the fact that both naturalists saw
in evolution and its causes the great questions of the
immediate future, and by the agreement in the
interpretations which they independently offered.
Wallace was collecting and observing in the Malay
Archipelago, and wrote to Darwin as the one man
most likely to sympathise with and understand his
views and to offer valuable criticism.

In the "Annals and Magazine of Natural History"
for 1855, Wallace published a paper "On the Law
that has Regulated the Introduction of New
Species," and in this and a letter written from the
Malay Archipelago Darwin recognised the similarity
of their views, although the completeness of this
agreement was to be brought before him with
startling force a year after his sympathetic reply,
written May 1st, 1857. He then wrote:—

"By your letter and even still more by your paper in the
Annals, a year or more ago, I can plainly see that we have
thought much alike and to a certain extent have come to
similar conclusions. In regard to the Paper in the Annals,
I agree to the truth of almost every word of your paper; and
I dare say that you will agree with me that it is very rare to

find oneself agreeing pretty closely with any theoretical paper; for it is lamentable how each man draws his own different conclusions from the very same facts."

On December 22nd he replied to another letter from Wallace, again expressing agreement with all his conclusions except that upon the supposed continental extension to oceanic islands, on which, alluding to his previous discussion, he says :—

"You will be glad to hear that neither Lyell nor Hooker thought much of my arguments. Nevertheless, for once in my life, I dare withstand the almost preternatural sagacity of Lyell."

And he concludes with the wish—

"May all your theories succeed, except that on Oceanic Islands, on which subject I will do battle to the death."

He also said, as regards Wallace's conclusions: "I believe I go much further than you; but it is too long a subject to enter on my speculative notions."

Finally, on June 18th, 1858, Darwin received from Wallace a manuscript essay bearing the title "On the Tendency of Varieties to depart indefinitely from the Original Type." Upon this essay he wanted Darwin's opinion, and asked him, if he thought well of it, to forward it to Lyell. Darwin was startled to find in the essay a complete account of his own views. That very day he wrote to Lyell, enclosing the essay. In the letter he said :—

"Your words have come true with a vengeance—that I should be forestalled. You said this, when I explained to you here very briefly my views of 'Natural Selection' depending on the struggle for existence. I never saw a more striking coincidence ; if Wallace had my MS. sketch written out in 1842, he could not have made a better short abstract! Even his terms now stand as heads of my chapters."

A few days later (June 25th) he again wrote to Lyell, saying—

"I should be extremely glad now to publish a sketch of my general views in about a dozen pages or so ; but I cannot persuade myself that I can do so honourably. Wallace says nothing about publication, and I enclose his letter. But as I had not intended to publish any sketch, can I do so honourably, because Wallace has sent me an outline of his doctrine ? I would far rather burn my whole book, than that he or any other man should think that I had behaved in a paltry spirit."

He also asked Lyell to send the letter on to Hooker, "for then I shall have the opinion of my two best and kindest friends." He was so much distressed at the idea of being unfair to Wallace that he wrote again the next day to put the case against himself in an even stronger light. This must have been one of the most trying times in Darwin's life, for, in addition to the cause of trouble and perplexity described above, one of his children died of scarlet fever, and there was the gravest fear lest the others should be attacked.

Thus appealed to, Lyell and Hooker took an extremely wise and fair course. They asked Darwin for an abstract of his work, and, accepting the whole responsibility, communicated it and Wallace's essay in a joint paper to the Linnean Society, giving an account of the circumstances of the case in a preface,

which took the form of a letter to the Secretary of
the Society. In this letter they introduced to the
Society "the results of the investigations of the
indefatigable naturalists, Mr. Charles Darwin and
Mr. Alfred Wallace."

"These gentlemen having, independently and unknown to
one another, conceived the same very ingenious theory to
account for the appearance and perpetuation of varieties and
of specific forms on our planet, may both fairly claim the
merit of being original thinkers in this important line of
enquiry; but neither of them having published his views,
though Mr. Darwin has for many years past been repeatedly
urged by us to do so, and both authors having now unre-
servedly placed their papers in our hands, we think it would
best promote the interests of science that a selection from
them should be laid before the Linnean Society."

After giving a list of these selections, they say of
Wallace's essay—

"This was written at Ternate * in February, 1858, for the
perusal of his friend and correspondent Mr. Darwin, and sent
to him with the expressed wish that it should be forwarded to
Sir Charles Lyell, if Mr. Darwin thought it sufficiently novel
and interesting. So highly did Mr. Darwin appreciate the
value of the views therein set forth, that he proposed, in a
letter to Sir Charles Lyell, to obtain Mr. Wallace's consent to
allow the Essay to be published as soon as possible. Of this
step we highly approved, provided Mr. Darwin did not withhold from the public, as he was strongly inclined to do (in
favour of Mr. Wallace), the memoir which he had himself
written on the same subject, and which, as before stated, one
of us had perused in 1844, and the contents of which we had

* My friend Mr. J. J. Walker, R.N., tells me that the house in
which Wallace lived in Ternate, and in which the essay was written,
is still pointed out by the natives as one of the features of the place.
It is, unfortunately, much dilapidated.

both of us been privy to for many years. On representing this to Mr. Darwin, he gave us permission to make what use we thought proper of his memoir, &c. ; and in adopting our present course, of presenting it to the Linnean Society, we have explained to him that we are not solely considering the relative claims to priority of himself and his friend, but the interests of science generally ; for we feel it to be desirable that views founded on a wide deduction from facts, and matured by years of reflection, should constitute at once a goal from which others may start, and that, while the scientific world is waiting for the appearance of Mr. Darwin's complete work, some of the leading results of his labours, as well as those of his able correspondent, should together be laid before the public."

The title of the joint paper was " On the Tendency of Species to form Varieties ; and on the Perpetuation of Varieties and Species by Natural Means of Selection." It was read July 1st, 1858.

CHAPTER IX.

DARWIN'S SECTION OF THE JOINT MEMOIR READ
BEFORE THE LINNEAN SOCIETY JULY 1, 1858.

THE first section of Darwin's communication consisted of extracts from the Second Chapter of the First Part of his manuscript essay of 1844. The Part was entitled "The Variation of Organic Beings under Domestication, and in their Natural State," and the Second Chapter was headed "On the Variation of Organic Beings in a State of Nature; on the Natural Means of Selection; on the Comparison of Domestic Races and True Species." The extracts first deal with the tendency towards rapid multiplication and the consequent struggle for life. The average constancy of the numbers of individuals is traced to the average constancy of the amount of food, "whereas the increase of all organisms tends to be geometrical." Practical illustrations are given in the enormous increase of the mice in La Plata during the drought which killed millions of cattle, and in the well-known and rapid increase of the animals and plants introduced by man into a new and favourable country.

The checks which operate when the country is stocked and the species reaches its average are most difficult to detect, but none the less certain. If any check is lightened in the case of any organism it will

E

at once tend to increase. "Nature may be compared to a surface on which rest ten thousand sharp wedges touching each other and driven inward by incessant blows." Darwin meant by this image to express that just as any single wedge would instantly rise above the rest when the blows on it were in any way lessened as compared with those on the other wedges, so it would be with the proportionate number of any species when the checks to which it is subjected are in any way relaxed.

If the external conditions alter, and the changes continue progressing, the inhabitants will be less well adapted than formerly. The changed conditions would act on the reproductive system and render the organisation plastic. Now, can it be doubted, from the struggle each individual has to obtain subsistence, that any minute variation in structure, habits, or instincts adapting that individual better to the new conditions would tell upon its vigour and health? "Yearly more are bred than can survive; the smallest grain in the balance, in the long run, must tell on which death shall fall, and which shall survive." If this went on for a thousand generations who will deny its effect "when we remember what, in a few years, Bakewell effected in cattle, and Western in sheep, by this identical principle of selection"?

He gives an imaginary example of a canine animal preying on rabbits and hares. If the rabbits, constituting its chief food, gradually became rarer, and the hares more plentiful, the animal would be driven to try and catch more hares, and hence would be selected in the direction of speed and sharp eyesight. " I can see no more reason to doubt that these cases in a thousand generations would produce a marked effect, and adapt the form of the fox or dog to the catching of hares instead of rabbits, than that greyhounds can be improved by selection and careful breeding." So also with plants having seeds with rather more down, leading to wider dissemination. Darwin here added this note: "I can see no more difficulty in this, than in the planter improving his varieties of the cotton plant. C. D. 1858."

Then follows a brief sketch of sexual selection and a comparison with natural selection, and the conclusion is reached—" this kind of selection, however, is less vigorous than the other; it does not require the death of the less successful, but gives to them fewer descendants. The struggle falls, moreover, at a time of year when food is generally abundant, and perhaps the effect chiefly produced would be the modification of the secondary sexual characters, which are not related to the power of obtaining food, or to defence from enemies, but to fighting with or rivalling other males."

The second section was entitled "Abstract of a Letter from C. Darwin, Esq., to Professor Asa Gray, Boston, U.S., dated Down, September 5th, 1857." To this letter Darwin attached great importance as a convenient and brief account of the essentials of his theory, written and sent to Asa Gray many months before he received Wallace's essay. A tolerably full abstract of the letter, which is itself a very brief abstract, is therefore printed below. The epitome here given is taken from the letter itself, and is in certain respects more full than that published in the Linnean Journal.

In the introductory parts Darwin explained that "the facts which kept me longest scientifically orthodox are those of adaptation—the pollen-masses in asclepias—the mistletoe, with its pollen carried by insects, and seed by birds—the woodpecker, with its feet and tail, beak and tongue, to climb the tree and secure insects. To talk of climate or Lamarckian habit producing such adaptations to other organic beings is futile. This difficulty I believe I have surmounted." Having then stated that the reasons which induced him to accept evolution were "general facts in the affinities, embryology, rudimentary organs, geological history, and geographical distribution of organic beings," he proceeds to give a brief account of his "notions on the means by which Nature makes her species." The following is an abstract of the account he gives:—

I. The success with which selection has been applied by man in making his breeds of domestic animals and plants : and this even in ancient times when the selection was unconscious, viz. when breeding was not thought of, but the most useful animals and plants were kept and the others destroyed. "Selection acts only by the accumulation of very slight or greater variations," and man in thus accumulating "*may be said* to make the wool of one sheep good for carpets, and another for cloth, &c."

II. Slight variations of all parts of the organism occur in nature, and if a being could select with reference to the whole structure, what changes might he not effect in the almost unlimited time of which geology assures us.

III. Animals increase so fast that, but for extermination, the earth would not hold the progeny of even the slowest breeding animal. Only a few in each generation can live ; hence the struggle for life, which has never yet been sufficiently appreciated. " What a trifling difference must often determine which shall survive and which perish ! " Thus is supplied the " unerring power " of " *Natural Selection* . . . which selects exclusively for the good of each organic being."

IV. If a country were changing the altered conditions would tend to cause variation, "not but what I believe most beings vary at all times enough for selection to act on." Extermination would expose the remainder to "the mutual action of a different set of inhabitants, which I believe to be more important to the life of each being than mere climate." In the infinite complexity of the struggle for life " I cannot doubt that during millions of generations individuals of a species will be born with some slight variation profitable to some part of its economy ; such will have a better chance of surviving and propagating this variation, which again will be slowly increased by the accumulative action of natural selection ; and the variety thus formed will either coexist with, or more commonly will exterminate its parent form." Thus complex adaptations like those of woodpecker or mistletoe may be produced.

V. Numerous difficulties can be answered satisfactorily in time. The supposed changes are only very gradual, and very slow, "only a few undergoing change at any one time." The imperfection of the geological record accounts for deficient direct evidence of change.

VI. Divergence during evolution will be an advantage. " The same spot will support more life if occupied by very diverse forms." Hence during the increase of species into its offspring—varieties, or sub-species, or true species, the latter " will try (only few will succeed) to seize on as many and as diverse places in the economy of nature as possible," and so will tend to "exterminate its less well-fitted parent." This explains classification, in which the organic beings "always *seem* to branch and sub-branch like a tree from a common trunk; the flourishing twigs destroying the less vigorous—the dead and lost branches rudely representing extinct genera and families."

In a postscript he says :—

"This little abstract touches only the accumulative power of natural selection, which I look at as by far the most important element in the production of new forms. The laws governing the incipient or primordial variation (unimportant except as the groundwork for selection to act on, in which respect it is all important), I shall discuss under several heads, but I can come, as you may well believe, only to very partial and imperfect conclusions."

It is, I think, of especial interest to find Darwin at this early period arguing in a most convincing manner for the creative power of natural selection. The selective power becomes, by accumulation, of such paramount importance in the process, as compared with the variations, that, although these latter are absolutely essential, man may be said to *make* his domestic breeds and Nature her species. The man who argued thus had been through and had left behind the difficulty that, even now, is often raised—that "before anything can be selected it must be," and therefore that selection is of small account as compared with variation.

CHAPTER X.

WALLACE'S SECTION OF THE JOINT MEMOIR READ
BEFORE THE LINNEAN SOCIETY JULY 1, 1858.

THE communication by Alfred Russel Wallace was entitled "On the Tendency of Varieties to depart indefinitely from the Original Type." An abstract of it is given below.

Varieties produced in domesticity are more or less unstable, and often tend to return to the parent form. This is usually thought to be true for all varieties, and to be a strong argument for the original and permanent distinctness of species.

On the other hand, races forming "permanent or true varieties" are well known, and there are generally no means of determining which is the *variety* and which the original *species*. The hypothesis of a "permanent invariability of species" is satisfied by supposing that, while such varieties cannot diverge from the species beyond a certain fixed limit, they may return to it.

This argument is founded on the assumption that *varieties* in nature are in all respects identical with those of, domestic animals. The object of the paper is to show that this is false, and "that there is a general principle in nature which will cause

many *varieties* to survive the parent species and
to give rise to successive variations departing further
and further from the original type." The same
principle explains the tendency of domestic animals
to return to the parent form.

"The life of wild animals is a struggle for
existence." To procure food and escape enemies
are the primary conditions of existence, and determine abundance and rarity, frequently seen in
closely allied species.

"Large animals cannot be so abundant as small
ones ; the carnivora must be less numerous than the
herbivora," eagles and lions than pigeons and ante-
lopes. Fecundity has little or nothing to do with
this. The least prolific animals would increase
rapidly if unchecked. But wild animals do not
increase beyond their average ; hence there must
be an immense amount of destruction. The abundance of species in individuals bears no relation
whatever to their fertility. Thus the excessively
abundant passenger pigeon of the United States
lays only one or two eggs. Its abundance is explained by the widespread supply of food rendered
available by its powers of flight. The food-supply
"is almost the sole condition requisite for ensuring
the rapid increase of a given species." This explains
why the sparrow is more abundant than the red-
breast, why aquatic species of birds are specially
numerous in individuals, why the wild cat is rarer
than the rabbit. "So long as a country remains

physically unchanged, the numbers of its animal population cannot materially increase." If one species does so, others must diminish. In the immense amount of destruction the weakest must die, "while those that prolong their existence can only be the most perfect in health and vigour—those who are best able to obtain food regularly and to avoid their numerous enemies. It is, as we commenced by remarking, 'a struggle for existence,' in which the weakest and least perfectly organised must always succumb."

This tendency must apply to species as well as individuals, the best adapted becoming abundant, the others scarce or even extinct. If we knew the whole of the conditions and powers of a species "we might be able even to calculate the proportionate abundance of individuals, which is the necessary result."

Hence, first, *the animal population of a country is generally stationary (due to food and other checks)*; second, *comparative abundance or scarcity of individuals is entirely due to organisation and resulting habits, the varying measure of success in the struggle being balanced by a varying population in a given area.*

Variations from type must nearly always affect habits or capacities. Even changes of colour may promote concealment, while changes in the limbs or any external organs would affect the mode of procuring food, etc. "An antelope with shorter or weaker legs must necessarily suffer more from the

attacks of the feline carnivora "; the passenger pigeon
with less powerful wings could not always procure
sufficient food. Hence species thus modified would
gradually diminish ; but, on the other hand, if
modified in the direction of increased powers, would
become more numerous. Varieties will fall under
these two classes—those which will never rival, and
those which will eventually outnumber, the parent
species. If, then, some alteration in conditions
occurred making existence more difficult to a certain
species, first the less favourable variety would suffer
and become extinct, then the parent species, while
the superior variety would alone remain, "and on a
return to favourable circumstances would rapidly
increase in numbers and occupy the place of the
extinct species and variety."

The superior *variety* would thus replace the
species, to which it *could not* return, for the latter
could never compete with the former. Hence a
tendency to revert would be checked. But the
superior variety, when established, would in time give
rise to new varieties, some of which would become
predominant. Hence *progression and continued
divergence* would follow, but not invariably, for the
criteria of success or failure would vary, and would
sometimes render a race which was under other
conditions the most favoured now the least so.
Variations without any effect on the life-preserving
powers might also occur. But it is contended that
certain varieties must, on the average, tend to persist

longer than the parent species, while the scale on which nature works is so vast that an average tendency must in the end attain its full result.

Comparing domestic with wild animals, the very existence of the latter depends upon their senses and physical powers. Not so with the former, which are defended and fed by man.

Any favourable variety of a domestic animal is utterly useless to itself; while any increase of the powers and faculties of wild animals is immediately available, creating, as it were, a new and superior animal.

Again, with domestic animals all variations have an equal chance, and those which would be extremely injurious in a wild state are, under the artificial conditions, no disadvantage. Our domestic breeds could never have come into existence in a wild state, and if turned wild "*must* return to something near the type of the original wild stock, *or become altogether extinct.*"*

Hence we cannot argue from domestic to wild animals, the conditions of life in the two being completely opposed.

Lamarck's hypothesis of change produced by the attempts of animals to increase the development of

---

* Wallace has added the following note to the reprint in "Natural Selection and Tropical Nature," London, 1891, p. 31: "That is, they will vary, and the variations which tend to adapt them to the wild state, and therefore approximate them to wild animals, will be preserved. Those individuals which do not vary sufficiently will perish."

their own organs has been often refuted, but the view
here proposed depends upon the action of principles
constantly working in nature.   Retractile talons of
falcons and cats have not been developed by volition,
but by *the survival of those which had the greatest
facilities for seizing prey.*   The long neck of the
giraffe was not produced by constant stretching, but
by the success which any increase in the length of
neck ensured to its possessors.   Even colours, especially of insects, are explained in the same way, for
among the varieties of many tints, those "having
colours best adapted to concealment . . . would in-
evitably survive the longest."   We can similarly
explain deficiency of some organs with compensating
development of others, "great velocity making up for
the absence of defensive weapons," etc.   Varieties
with an unbalanced deficiency could not long survive.
The action of the principle is like the governor of a
steam-engine, checking irregularities almost before
they become evident.   Such a view accords well with
" the many lines of divergence from a central type";
the increasing efficiency of a particular organ in a
series of allied species ; the persistence of unimportant
parts when important ones have changed; the "more
specialised structure," said by Owen to be characteristic of recent as compared with extinct forms.

Hence there is a tendency of certain classes of
varieties to progress further and further from the
original type, and there is no reason for assigning any
limit to this progression.   Such gradual changes

" may, it is believed, be followed out so as to agree with all the phenomena presented by organised beings, their extinction and succession in past ages, and all the extraordinary modifications of form, instinct, and habits which they exhibit."

Wallace's Essay has been reprinted without alteration in his "Essays on Natural Selection," recently re-issued combined with "Tropical Nature."

CHAPTER XI.

COMPARISON OF DARWIN'S AND WALLACE'S SECTIONS
OF THE JOINT MEMOIR—RECEPTION OF THEIR
VIEWS—THEIR FRIENDSHIP.

COMPARING the essays of these two naturalists, we observe that Darwin here first makes public the phrase "natural selection," Wallace the "struggle for existence"; although so closely do their lines of thought converge that Darwin, using practically the same words, speaks of the "struggle for life." Both show, by examples, the tendency of all animals to multiply at an enormous rate, and both show that their tolerably constant numbers are due to the constant supply of food.

Both treat of domesticated animals, but in very different ways. Darwin uses them as the practical illustration of selection, and argues that if man by selection can make such forms, Nature can make her species by the same means. Wallace disposes of the argument that the reversion of domesticated varieties to the wild form is a proof of the permanent distinctness of species, by showing in some detail that the former are "abnormal, irregular, artificial."

Neither of them draws any distinction between instinct and other qualities, but assumes that the former is, like the latter, operated upon by natural selection.

Wallace makes a special point of protective resemblances in the colours of insects, etc.

The important principle of " divergence of character," and the relatively unimportant one of " sexual selection," are both clearly explained by Darwin.

Neither writer speaks of the direct effect of external conditions—except as a cause of plasticity by Darwin—or the inherited effects of use and disuse. Lamarck is mentioned only to be dismissed by Wallace. The evolution of the giraffe's long neck is explained by Wallace on the principle of natural selection, which is contrasted with Lamarck's original explanation of the same character. This contrast, which has been so often drawn, was therefore originally contained in the first public statement of natural selection.

As has been indicated above, Darwin suggested a cause of variation in the direct effect of changed external conditions on the reproductive system.

In comparing the two essays it is not unnatural to conclude, as Professor Osborn has done (" From the Greeks to Darwin," 1894, p. 245), that the two writers held different views upon the material utilised by natural selection in the production of new species, Darwin relying upon the usual slight differences which separate individuals and upon variations in single characters, Wallace upon fully formed varieties—viz. individuals which departed

conspicuously from the type of the species, and
which may exist singly or in considerable numbers
side by side with the parent form.

Professor Osborn's actual words are as follows:—

"Darwin dwells upon *variations in single characters*, as
taken hold of by Selection ; Wallace mentions variations, but
dwells upon *full-formed varieties*, as favourably or unfavour-
ably adapted. It is perfectly clear that with Darwin the
struggle is so intense that the chance of survival of each
individual turns upon a single and even slight variation.
With Wallace, varieties are already presupposed by causes
which he does not discuss, a change in the environment occurs,
and those varieties which happen to be adapted to it survive.
There is really a wide gap between these two statements and
applications of the theory."

Further consideration tends to obliterate this
supposed distinction. Although Wallace used the
term "variety" as contrasted with "species," the
whole context proves that he, equally with Darwin,
recognised the importance of individual variations
and of variations in single characters. This becomes
clear when we remember his argument about the
neck of the giraffe, the changes of colour and hairi-
ness, the shorter legs of the antelope, and the less
powerful wings of the passenger pigeon. Wallace
has kindly written to me (May 12th, 1896) stating
the case as I have given it, and he further explains—

"I used the term 'varieties' because 'varieties' were alone
recognised at that time, individ[l] variability being ignored
or thought of *no importance*. My 'varieties' therefore included 'individual variations.'"

On the other hand, Darwin certainly included large single variations (in other words, " varieties ") as well as ordinary individual differences, among the material for natural selection, and he did not abandon the former until he was convinced by the powerful reasoning of Fleeming Jenkin (*North British Review*, June, 1867), who argued that single large differences of a sudden and conspicuous kind (Darwin's " variations ") would certainly be swamped by intercrossing. Upon this review of the " Origin " Francis Darwin says (" Life and Letters ")—

" It is not a little remarkable that the criticisms which my father, as I believe, felt to be the most valuable ever made on his views should have come, not from a professed naturalist but from a Professor of Engineering."

After reading this review, Darwin wrote to Wallace (January 22nd, 1869):—

" I always thought individual differences more important than single variations, but now I have come to the conclusion that they are of paramount importance, and in this I believe I agree with you. Fleeming Jenkin's arguments have convinced me."

The ambiguity of this sentence evidently misled Wallace into believing that the single variations were considered of paramount importance. Darwin therefore wrote again (February 2nd):—

" I must have expressed myself atrociously ; I meant to say exactly the reverse of what you have understood. F. Jenkin argued in the 'North British Review' (June 1867) against

F

single variations ever being perpetuated, and has convinced me, though not in quite so broad a manner as here put. I always thought individual differences more important ; but I was blind and thought single variations might be preserved much oftener than I now see is possible or probable. I mentioned this in my former note merely because I believed that you had come to a similar conclusion, and I like much to be in accord with you. I believe I was mainly deceived by single variations offering such simple illustrations, as when man selects."

From these two letters to Wallace we see that the latter was the first to give up the larger variations in favour of ordinary individual differences.

Darwin also wrote to Victor Carus on May 4th, 1869 :—

"I have been led to . . . infer that single variations are even of less importance, in comparison with individual differences, than I formerly thought."

There has been much misconception on this point, and a theory of evolution by the selection of large single variations—a view held by many, but not by Darwin—has been passed off as the Darwinian theory of natural selection. It is surprising that this old mistake should have been repeated at so recent a date, and on so important an occasion as the Presidential Address to the British Association at Oxford on August 8th, 1894, and that so ill-aimed a criticism should have been quoted with approval in a leading article in the *Times* of the following day. The following extracts from Lord Salisbury's address unfortunately leave no doubt on the matter :

" What is to secure that the two individuals of opposite sexes
in the primeval forest, who have been both accidentally blessed
with the same advantageous variation shall meet, and transmit
by inheritance that variation to their successors ? . . . The biologists do well to ask for an immeasurable expanse of time, if
the occasional meetings of advantageously varied couples from
age to age are to provide the pedigree of modifications which
unite us to our ancestor the jelly-fish. . . . There would be
nothing but mere chance to secure that the advantageously
varied bridegroom at one end of the wood should meet the
bride, who by a happy contingency had been advantageously
varied in the same direction at the same time at the other end
of the wood. It would be a mere chance if they ever knew of
each other's existence—a still more unlikely chance that they
should resist on both sides all temptations to a less advantageous alliance. But unless they did so, the new breed would
never even begin, let alone the question of its perpetuation
after it had begun."

It is of interest to reproduce Lord Salisbury's
words in close proximity to Darwin's real statements on the subject, as shown in the letters to
his friends—statements which are also expressed in
many places in his published works.

The joint paper was read before the Linnean
Society on July 1st, 1858, about a fortnight
after Wallace's essay had been received by Darwin.
There was no discussion, but the interest and
excitement at the meeting were very great,
owing in large part to the influential support with
which the new theory came before the scientific
world. Darwin appreciated the importance of this
support at its true value, for he wrote to Hooker,
July 5th :—

" You must know that I look at it, as very important, for
the reception of the view of species not being immutable, the
fact of the greatest Geologist and Botanist in England taking
any sort of interest in the subject : I am sure it will do much
to break down prejudices."

In the following January Darwin received a letter
from Wallace, and his reply (on the 25th) shows how
much relieved and pleased he was at its generous
spirit. Alluding to Lyell's and Hooker's action in his
" Autobiography " Darwin says :—" I was at first very
unwilling to consent, as I thought Mr. Wallace
might consider my doing so unjustifiable, for I did
not then know how generous and noble was his
disposition." It was this letter which conveyed the
knowledge to him and set his mind at rest on the
subject.

Thus ended one of the most interesting and
memorable episodes in the history of science. It was
sufficiently remarkable that two naturalists in widely-
separated lands should have independently arrived
at the theory which was to be the turning-point in
the history of biology and of many other sciences—
although such simultaneous discoveries have been
known before; it was still more remarkable that one
of the two should unknowingly have chosen the
other to advise him upon the theory which was to
be for ever associated with both their names. It
was a magnificent answer to those who believed that
the progress of scientific discovery implies continual
jealousy and bitterness, that the conditions attending

the first publication of the theory of natural selection were the beginning of a life-long friendship and of mutual confidence and esteem.*

It is justifiable to speak of this episode as the *beginning* of Darwin's and Wallace's friendship, for the latter writes (February, 1895):—

"I had met him *once* only for a few minutes at the Brit. Mus. before I went to the East."

Later on Darwin, in his letters to Wallace, more than once alluded to the simultaneous publication of their essays. Thus he wrote, April 18th, 1869, congratulating Wallace on his article in the *Quarterly Review* for that month:—

"I was also much pleased at your discussing the difference between our views and Lamarck's. One sometimes sees the odious expression, 'Justice to myself compels me to say,' &c., but you are the only man I ever heard of who persistently does himself an injustice, and never demands justice. Indeed, you ought in the review to have alluded to your paper in the 'Linnean Journal,' and I feel sure all our friends will agree in this. But you cannot 'Burke' yourself however much you may try, as may be seen in half the articles which appear."

* Since the above paragraph was written I have again read Professor Newton's eloquent Address to the Biological Section of the British Association at Manchester in 1887, and find that he says on the same subject—"If in future you should meet with any cynic who may point the finger of scorn at the petty quarrels in which naturalists unfortunately at times engage, particularly in regard to the priority of their discoveries, you can always refer him to this greatest of all cases, where scientific rivalry not only did not interfere with, but even strengthened, the good-feeling which existed between two of the most original investigators" (Report of Meeting, p. 731).

And again, on April 20th of the following year, he wrote :—

"I hope it is a satisfaction to you to reflect—and very few things in my life have been more satisfactory to me—that we have never felt any jealousy towards each other, though in one sense rivals. I believe that I can say this of myself with truth, and I am absolutely sure that it is true of you."

CHAPTER XII.

WE have already seen in the earlier part of this volume, the gradual development of the theory of Natural Selection in the mind of Darwin, and the long succession of experiments and observations which he undertook before he could bring himself to publish anything upon the subject, as well as the conditions which forced him to a hurried publication in the end. It is of the deepest interest to compare with this the account which Wallace has given us of the mental process by which he arrived at the same conclusions.

This deeply interesting personal history has only been known during the last few years ; in 1891 Wallace republished his "Essays on Natural Selection" in one volume, combined with "Tropical Nature," and he has added (on pp. 20, 21) the following introductory note to Chapter II., viz. the reprint of his Linnean Society Memoir "On the Tendencies of Varieties to depart indefinitely from the Original Type." The note is here reprinted in full :—

"As this chapter sets forth the main features of a theory identical with that discovered by Mr. Darwin many years

before but not then published, and as it has thus an
historical interest, a few words of personal statement may
be permissible. After writing the preceding paper ["On
the Law which has Regulated the Introduction of New
Species"] the question of *how* changes of species could have
been brought about was rarely out of my mind, but no
satisfactory conclusion was reached till February 1858. At
that time I was suffering from a rather severe attack of
intermittent fever at Ternate in the Moluccas, and one day,
while lying on my bed during the cold fit, wrapped in
blankets, though the thermometer was at 88° Fahr., the
problem again presented itself to me, and something led me to
think of the 'positive checks' described by Malthus in his
'Essay on Population,' a work I had read several years
before, and which had made a deep and permanent impression
on my mind. These checks—war, disease, famine and the like—
must, it occurred to me, act on animals as well as man. Then
I thought of the enormously rapid multiplication of animals,
causing these checks to be much more effective in them than
in the case of man; and while pondering vaguely on this fact
there suddenly flashed upon me the *idea* of the survival of the
fittest—that the individuals removed by these checks must be
on the whole inferior to those that survived. In the two hours
that elapsed before my ague fit was over I had thought out
almost the whole of the theory, and the same evening I
sketched the draft of my paper, and in the two succeeding
evenings wrote it out in full, and sent it by the next post to
Mr. Darwin. Up to this time the only letters I had received
from him were those printed in the second volume of his
Life and Letters (vol. ii., pp. 95 and 108), in which he speaks
of its being the twentieth year since he 'opened his first
note-book on the question how and what way do species and
varieties differ from each other,' and after referring to oceanic
islands, the means of distribution of land-shells, &c., added:
'My work, on which I have now been at work more or less for
twenty years, *will not fix or settle anything*; but I hope it
will aid by giving a large collection of facts, with one definite
end.' The words I have italicised, and the whole tone of his
letters, led me to conclude that he had arrived at no definite
view as to the origin of species, and I fully anticipated that

my theory would be new to him, because it seemed to me to settle a great deal. The immediate result of my paper was that Darwin was induced at once to prepare for publication his book on the *Origin of Species* in the condensed form in which it appeared, instead of waiting an indefinite number of years to complete a work on a much larger scale which he had partly written, but which in all probability would not have carried conviction to so many persons in so short a time. I feel much satisfaction in having thus aided in bringing about the publication of this celebrated book, and with the ample recognition by Darwin himself of my independent discovery of 'natural selection.' (See *Origin of Species*, 6th ed., introduction, p. 1, and Life and Letters, vol. ii., chap. iv., pp. 115-129 and 145)."

A very similar account, differing in a few unimportant details from that quoted above, was written December 3rd, 1887, by Wallace to Professor Newton, and is published in the abridged "Life and Letters of Charles Darwin" (1892; pp. 189, 190). At the conclusion Wallace says:—

". . . . I *had* the idea of working it out, so far as I was able, when I returned home, not at all expecting that Darwin had so long anticipated me. I can truly say *now*, as I said many years ago, that I am glad it was so ; for I have not the love of *work*, *experiment* and *detail* that was so pre-eminent in Darwin, and without which anything I could have written would never have convinced the world."

It is of great interest to learn that Wallace as well as Darwin was directed to natural selection by Malthus Essay. Hence, as the late Professor Milnes Marshall has pointed out (Lectures on the Darwinian Theory, pp. 212, 213), the laws of the multiplication and extinction of man suggested to both naturalists

those more general laws by which it was possible to
understand the development of the whole animal and
vegetable worlds.

There is a tremendous contrast between these two
discoverers, in the speed with which they respectively
developed their ideas on the subject into a shape
which satisfied them as suitable for publication.
Wallace, after the inspiration which followed his
reflections upon Malthus, had "thought out almost
the whole of the theory" in two hours, and in three
evenings had completed his essay. Darwin, receiving
the same inspiration from the same source, in October
1838, wrote a brief account of it after four years'
reflection and work, and finished a longer account two
years later, but was not prepared to give anything to
the public until he was compelled to do so fourteen
years later in 1858. All this delay was of the greatest
advantage when a full exposition of the theory finally
came before the world in the " Origin of Species"; for
all difficulties had been fully considered and answered
beforehand, while the wealth of new facts by which it
was supported compelled a respectful hearing for the
theory itself.

Wallace, like Darwin, was convinced of evolution
before he discovered any principle which supplied a
motive cause for the process. This conviction is
expressed very clearly in his interesting essay already
alluded to " On the Law which has regulated the
Introduction of New Species" (Ann. and Mag., Nat.
Hist., 1855, p. 184; reprinted without alteration in his

Essays on Natural Selection). The law he states in these words:—

"Every species has come into existence coincident both in time and space with a pre-existing closely allied species,"

a law which, as he justly claims for it,

"connects together and renders intelligible a vast number of independent and hitherto unexplained facts. The natural system of arrangement of organic beings, their geographical distribution, their geological sequence, the phenomena of representative and substituted groups in all their modifications, and the most singular peculiarities of anatomical structure, are all explained and illustrated by it, in perfect accordance with the vast mass of facts which the researches of modern naturalists have brought together, and, it is believed, not materially opposed to any of them. It also claims a superiority over previous hypotheses, on the ground that it not merely explains, but necessitates what exists. Granted the law, and many of the most important facts in Nature could not have been otherwise, but are almost as necessary deductions from it, as are the elliptic orbits of the planets from the law of gravitation."

This important essay is dated by Wallace from Sarawak, Borneo, February, 1855.

The conclusions remind us of the words Darwin wrote in his note-book in 1837. "Led to comprehend true affinities. My theory would give zest to recent and Fossil comparative Anatomy." By his theory Darwin here means evolution and not natural selection, which was not discovered by him until the end of 1838.

CHAPTER XIII.

CANON TRISTRAM THE FIRST PUBLICLY TO ACCEPT THE THEORY OF NATURAL SELECTION (1859).

ALTHOUGH the historic meeting at the Linnean Society appeared to produce but little effect, one distinguished naturalist publicly accepted the theory of natural selection before the publication of "The Origin of Species," and therefore as the direct result of Darwin's and Wallace's joint paper. This great distinction belongs to Canon Tristram, as Professor Newton has pointed out in his Presidential Address to the Biological Section of the British Association at Manchester in 1887 ("Reports," p. 727), at the same time expressing the hope "that thereby the study of Ornithology may be said to have been lifted above its fellows."

Canon Tristram's paper, "On the Ornithology of Northern Africa" (Part iii., The Sahara, continued), was published in *The Ibis*, vol. i., October, 1859. The important conclusions alluded to above are contained at the end of the section upon the species of desert larks (pp. 429–433):

"Writing with a series of about 100 larks of various species from the Sahara before me, I cannot help feeling convinced of the truth of the views set forth by Messrs. Darwin and Wallace in their communications to the Linnean Society, to which my friend Mr. A. Newton last year directed my attention. . . . It is hardly possible, I should think, to illustrate this theory better than by the larks and chats of North Africa."

In all these birds we trace gradual modifications of coloration and of anatomical structure, deflecting by very gentle gradations from the ordinary type ; but when we take the extremes, presenting most marked differences."

These differences, he concludes—

"have a very direct bearing on the ease or difficulty with which the animal contrives to maintain its existence."

He then points out, upon the uniform surface of the desert it is absolutely necessary that animals shall be protected by their colour:

"Hence, without exception, the upper plumage of every bird, whether Lark, Chat, Sylvian, or Sandgrouse, and also the fur of all the small mammals, and the skin of all the Snakes and Lizards, is of one uniform isabelline or sand colour. It is very possible that some further purpose may be served by the prevailing colours, but this appears of itself a sufficient explanation. There are individual varieties in depth of hue among all creatures. In the struggle for life which we know to be going on among all species, a very slight change for the better, such as improved means of escaping from its natural enemies (which would be the effect of an alteration from a conspicuous colour to one resembling the hue of the surrounding objects), would give the variety that possessed it a decided advantage over the typical or other forms of the species. Now in all creatures, from Man downwards, we find a tendency to transmit individual varieties or peculiarities to the descendants. A peculiarity either of colour or form soon becomes hereditary when there are no counteracting causes, either from change of climate or admixture of other blood. Suppose this transmitted peculiarity to continue for some generations, especially when manifest advantages arise from its possession, and the variety becomes not only a race, with its variations still more strongly imprinted upon it, but it becomes the typical form of that country."

Canon Tristram then points out the manner in which he imagines that one of the crested larks of the desert has been produced by the survival of the lightest coloured individuals, *Galerida abyssinica* only differing in this respect from *G. cristata* of Europe. Short-billed species of the same genus inhabiting hard rocky districts, and long-billed inhabiting loose sandy tracts have, he believes, been produced by the survival in each case of the forms of bill most suited to procure food:

"Here are only two causes enumerated which might serve to *create* as it were a new species from an old one, yet they are perfectly natural causes, and such as, I think, must have occurred, and are possibly occurring still. We know so very little of the causes which in the majority of cases make species rare or common, that there may be hundreds of others at work, some even more powerful than these, which go to perpetuate and eliminate certain forms 'according to natural means of selection.' But even these superficial causes appear sufficient to explain the marked features of the Desert races, which frequently approach so very closely the typical form, and yet possess such invariably distinctive characteristics, that naturalists seem agreed to elevate them to the rank of species."

Although the author also declares his belief in the special creation of many species—a view put forward as possible by Darwin in the "Origin"*—and also believed in some direct influence of locality, climate, etc., the above quoted passages are a most complete acceptance of natural selection, at the same time affording excellent examples of its operation.

* "There is grandeur in this view of life, with its several powers, having been originally breathed by the Creator into a few forms or into one; . . ."—(Concluding paragraph of "Origin," 1860, p. 490.)

CHAPTER XIV.

Almost immediately after the Linnean Society meeting, and evidently earlier than September, the time mentioned in his "Autobiography," Darwin began to prepare a longer and more complete account of his work on evolution and natural selection. This account was at first intended for the Linnean Society, but it was soon found to be too long, and he then decided to publish it as an independent volume. In thus preparing the manuscript for what afterwards became the "Origin of Species," Darwin tells us ("Autobiography") he acted under "the strong advice of Lyell and Hooker," and his letters also show the great interest that they were taking in the work.

Darwin seems to have found the "Origin"—or his "Abstract," as he always calls it—very hard work, and he ends his letter to Wallace (January 25th, 1859) with the words:

"I look at my own career as nearly run out. If I can publish my Abstract and perhaps my greater work on the same subject, I shall look at my course as done."

At the same time, so great was his enthusiasm and interest, in spite of the hard work and ill-health, that all through this period he was making fresh observations whenever an opportunity occurred. Thus we find him writing to Hooker about the thistle-down blown out to sea and then back to shore again; about the migrations of slave-making ants which he had been watching; about the bending of the pistil into the line of the gangway leading to the honey when this latter "is secreted at one point of the circle of the corolla," etc. And on March 2nd, 1859, he writes about "an odd, though very little, fact":—Large nuts had been found in the crops of some nestling Petrels at St. Kilda, which he suspected the parent birds had picked up from the Gulf Stream. He arranged for one of these to be sent, and asked Hooker for the name and country. He asks forgiveness for the trouble, "for it is a funny little fact after my own heart." The nuts turned out to be West Indian.

When the proposal for publication had been accepted by Murray and the manuscript was assuming its final form, the letters to Hooker were more frequent than ever. Writing on May 11th, 1859, Darwin again raises the question of the relative importance of variation and selection.

"I imagine from some expressions . . . that you look at variability as some necessary contingency with organisms, and further that there is some necessary tendency in the variability to go on diverging in character or degree. *If you do*, I do not agree."

Darwin's splendid confidence in the future appears in a letter written about this time (September 2, 1859) in which he begs Lyell not to commit himself "to go a certain length and no further; for," he says, "I am deeply convinced that it is absolutely necessary to go the whole vast length, or stick to the creation of each separate species." He asks Lyell to remember that his verdict will probably be of more importance than the book itself in influencing the present acceptance or rejection of the views. "In the future," he continues, "I cannot doubt about their admittance, and our posterity will marvel as much about the current belief as we do about fossil shells having been thought to have been created as we now see them." And again writing to Lyell a few days later (September 20th), he says, "I cannot too strongly express my conviction of the general truth of my doctrines, and God knows I have never shirked a difficulty."

I have thought it well to bring strong evidence of Darwin's entire confidence in his conclusions, because his writings were so extraordinarily balanced and judicial, and the weight he gives to opposing considerations so great, that a superficial student might imagine that he wrote and argued without any very strong convictions.

The letters to Mr. John Murray, the publisher, are eminently characteristic, in the expressions of regret for trouble given, and of pleasure at the work done, in the scrupulous care to prevent the publisher from feeling committed, if on further acquaintance with the manuscript he did not wish to accept it, and in the offer to contribute towards the cost of corrections.

G

The first edition of "The Origin of Species" was published November 24th, 1859. The edition consisted of 1,250 copies, all of which were sold on the day of issue.

The full title of this volume, of which Darwin justly says ("Autobiography"), "It is no doubt the chief work of my life," is reproduced below.

ON

THE ORIGIN OF SPECIES

BY MEANS OF NATURAL SELECTION,

OR THE

PRESERVATION OF FAVOURED RACES IN THE STRUGGLE FOR LIFE.

By CHARLES DARWIN, M.A.,

FELLOW OF THE ROYAL, GEOLOGICAL, LINNEAN, ETC., SOCIETIES ;
AUTHOR OF "JOURNAL OF RESEARCHES DURING H.M.S. 'BEAGLE'S' VOYAGE
ROUND THE WORLD."

This title is of interest, as has been pointed out by Professor E. Ray Lankester, in relation to the controversy upon the exact meaning of the word " Darwinism." Some writers have argued that the term " Darwinism " includes the whole of the causes of evolution accepted by Darwin—the supposed inherited effects of use and disuse and the direct influence of environment, which find a subordinate place in the " Origin," as well as natural selection, which is the real subject of the book and which is fully defined in the title. It would seem appropriate to use the term " Darwinism," as Wallace uses it, to indicate the causes of evolution which were suggested by Darwin himself, excluding those supposed causes which had been previously brought forward by earlier writers, and especially by Lamarck. The causes of evolution proposed by Lamarck are seriously disputed, and it is possible that they may be ultimately abandoned. If so, the integrity of " Darwinism," as interpreted by some controversialists, would be impaired ; and this, it will be generally admitted, would be most unfortunate, as well as most unfair to the memory of Darwin.

CHAPTER XV.

THE ORIGIN OF SPECIES (1859).

IT is very interesting to separate the two arguments which occur interwoven in the "Origin"—the argument for evolution and the argument for natural selection.   The paramount importance of Darwin's contributions to the evidences of organic evolution are often forgotten in the brilliant theory which he believed to supply the motive cause of descent with modification.   Organic evolution had been held to be true by certain thinkers during many centuries; but not only were its adherents entirely without a sufficient motive cause, but their evidences of the process itself were erroneous or extremely scanty.   It was Darwin who first brought together a great body of scientific evidence which placed the process of evolution beyond dispute, whatever the causes of evolution may have been.   And accordingly we find that, even at first, natural selection was attacked far more generally than the doctrine of descent with modification.

In Chapter I., Variation under Domestication and man's power of selection in forming breeds of animals and plants are discussed; in Chapter II., Variation under Nature; in Chapter III., the Struggle for

Existence ; in Chapter IV., which Darwin, in writing to his publisher, called " the Keystone of my Arch," the three preceding chapters are carried to their conclusion, and the operation of natural selection is explained and discussed. Hence, these four chapters deal almost exclusively with this process.

Chapter V. has for its subject the Laws of Variation, and explains causes of modification (external conditions, use and disuse, correlation, reversion, etc.) other than natural selection and the relation of the latter to the former.

In Chapter VI. difficulties are considered—partly those in the way of a belief in evolution and partly those which, at first sight, seem to be incapable of explanation on the theory of natural selection. Chapter VII. deals with a special difficulty of the latter kind, viz. Instinct, and shows how it can be accounted for by natural selection acting upon variation, although allowing some weight to the inheritance of habit. Chapter VIII. deals with Hybridism, the sterility of first crosses and of hybrids being considered as an objection to the doctrine of Descent with Modification. Chapter IX. treats of the Imperfection of the Geological Record as the explanation of the apparently insufficient evidence of evolution during past ages. Chapter X., on the Geological succession of Organic Beings, shows that, allowing for this Imperfection of Record, the facts brought to light by Geology support a belief in evolution and in some cases even in natural selection. Hence these five chapters deal partly with difficulties in the way of an acceptance of organic evolution and partly with those encountered by natural selection.

Of the three remaining chapters before the XIVth, and last, which contains the Recapitulation and Conclusion, two—XI. and XII.—are concerned with Geographical distribution, while the XIIIth deals with Classification, Morphology, Embryology and Rudimentary Organs. These three chapters are almost entirely devoted to the proof that the facts of Nature with which they deal are not inconsistent with, but rather support, and often strongly support, a belief in Organic Evolution.

Hence we see that this, incomparably the greatest work which the biological sciences have seen, begins with an explanation and defence and definition of the sphere of natural selection—then passes to consider difficulties which are partly those of natural selection, and partly of organic evolution—while it finally treats of the evidences of the latter process and the difficulties which a belief in it encounters.

This arrangement was a very wise one for a book which was intended to convince a large circle of readers; for the human mind so craves after an explanation, that it was of more importance for the success of the work to show first that an intelligible cause of evolution had been proposed, than to follow the more logical order of first setting forth the evidences of evolution.

The second edition (fifth thousand) was issued in

January, 1860, the third (seventh thousand) in 1861, the fourth (eighth thousand) in 1866, the fifth (tenth thousand) in 1869, the sixth in 1872: in 1887 the twenty-fourth thousand was reached.

A note to the last edition states that the second "was little more than a reprint of the first. The third edition was largely corrected and added to, and the fourth and fifth still more largely." The sixth edition also contains "numerous small corrections," and is about one-fourth larger than the first edition, although this material is, owing to the smaller print and more crowded lines, compressed into a smaller number of pages. The sixth edition also differs from the first in containing a glossary, an historical sketch, and a note and list of the chief corrections.

The titles of Chapters I., II., and III. remain the same in the first and last editions. Herbert Spencer's phrase is added to Darwin's term, as the heading of Chapter IV., which accordingly becomes in the last edition "Natural Selection; or the Survival of the Fittest." This change was certainly introduced in order to help readers to grasp the meaning of Darwin's title, which had been very generally misunderstood. The heading of Chapter V. remains the same, while in Chapter VI.—"Difficulties on Theory"—"on" is replaced by "of the." This chapter is, in the last edition, succeeded by a new one dealing with many of the difficulties which had been raised or had occurred to Darwin in the interval between the two editions; it is headed "Miscellaneous Objections to the Theory of Natural Selection." The titles of the remaining eight chapters are unchanged.

The first part of the title of the first edition— "On the Origin of Species"—becomes "The Origin of Species" in the last edition, and is still further shortened to "Origin of Species" on the outside of the volume.

The form of the earlier editions was admirably suited for the purpose of attracting, and—so far as was possible with so difficult a subject—convincing, a large number of readers. When the subject was new and strange, the more numerous details of the last edition, and the smaller print which became necessary, would have acted as a hindrance to the complete success of the work. Authors and publishers are sometimes apt to forget that the form of a book has a great deal to do with the absorption of the ideas contained in it, especially when the argument is from the nature of the case difficult to follow, and the subject a new one. Francis Darwin in the "Life and Letters" justly condemns the unattractive form of the sixth edition of the work.

CHAPTER XVI.

IN considering the reception of the "Origin of Species," it will be well first to show its effect upon Darwin's intimate scientific friends, most of whom had been familiar with his work for many years, and then to deal with its effects upon biologists generally, especially those of Darwin's own country.

The gradual strengthening of Darwin's influence over his old teacher Lyell, is one of the most interesting episodes in the personal history of the scientific men of this century.

Lyell, after reading the proof-sheets of the "Origin," wrote on October 3rd, 1859, praising the work very warmly, and suggesting a few improvements, some of which were adopted. Lyell hesitated to accept the theory, because he saw clearly that it would be impossible to stop short at the human species, while a common origin of men and beasts was distasteful to him. Thus, he said :—

"I have long seen most clearly that if any concession is made, all that you claim in your concluding pages will follow. It is this which has made me so long hesitate, always feeling that the case of man and his races, and of other animals, and that of plants is one and the same, and that if a ' vera causa ' be admitted for one, instead of a purely unknown and imaginary one, such as the word 'Creation,' all the consequences must follow."

To this letter Darwin replied (October 11th) at great length, in a most instructive letter, arguing in considerable detail on all the points alluded to by Lyell. He evidently thought that Lyell's opinion was of the utmost importance for the success of Natural Selection. "If ever you are [perverted]," he wrote at the end of the letter, "I shall know that the theory of Natural Selection is, in the main, safe."

About this time Darwin seems to have heard that Lyell had made up his mind to admit the doctrine of evolution into a new edition of the "Manual," and he wrote (November 23rd):—

"I honour you most sincerely. To have maintained in the position of a master, one side of a question for thirty years, and then deliberately give it up, is a fact to which I much doubt whether the records of science offer a parallel."

Lyell's public confession of faith was, however, not to be made for some years, and Darwin's letter was a little premature.

Space will not permit me to quote from the long correspondence with Lyell in the years following the appearance of the "Origin," although these letters are of the deepest interest, and deal in the most luminous manner with the difficulties of natural selection and evolution, as they appeared to one of the acutest intellects of that time. The letters soon began to produce an effect, and Darwin wrote (September 26th, 1860) to Asa Gray :—

"I can perceive in my immense correspondence with Lyell, who objected to much at first, that he has, perhaps unconsciously to himself, converted himself very much during the last six months, and I think this is the case even with Hooker. This fact gives me far more confidence than any other fact."

Later on Darwin evidently became a little annoyed that Lyell still delayed to declare his belief one way or the other. Thus he wrote to Asa Gray (May 11th, 1863) :—

"You speak of Lyell as a judge ; now what I complain of is that he declines to be a judge. . . I have sometimes almost wished that Lyell had pronounced against me. When I say 'me,' I only mean *change of species by descent.* That seems to me the turning-point. Personally, of course, I care much about Natural Selection ; but that seems to me utterly unimportant, compared to the question of Creation *or* Modification."

Shortly before this date, on February 24th, he wrote to Hooker in much the same style. These communications were called forth by the appearance of " The Antiquity of Man," and it is clear that Darwin's disappointment at Lyell's suspended judgment was due to their correspondence, which had encouraged him to expect some definite opinion on the question. "From all my communications with him, I must ever think that he has really entirely lost faith in the immortality of species," he wrote in his letter to Hooker. On March 6th he wrote to Lyell himself, expressing his disappointment, and again a few days later, rather complaining that his work was treated as a modification of Lamarck's :—

"This way of putting the case . . . closely connects Wallace's and my views with what I consider, after two deliberate readings, as a wretched book, and one from which (I well remember my surprise) I gained nothing."

When the second edition of "The Antiquity of Man" appeared in a few months, there was a significant change in one sentence :—

"Yet we ought by no means to undervalue the importance of the step which will have been made, should it hereafter become the generally received opinion of men of science (as I fully expect it will) that the past changes of the organic world have been brought about by the subordinate agency of such causes as Variation and Natural Selection."

The words in parentheses had been added, and constituted Lyell's first public expression of an opinion in favour of Darwin's views.

About this time an article appeared in the *Athenæum* (March 28th, 1863), attacking the opinions in favour of evolution contained in Dr. Carpenter's work on Foraminifera, and supporting spontaneous generation. This was one of the rare occasions on which Darwin entered into controversy, and he wrote attacking spontaneous generation, and pointing out the numerous classes of facts which are connected by an intelligible thread of reasoning by means of his theory. In this letter he quoted the altered sentence from the second edition of the "Antiquity." Darwin's letter was answered in an article (May 2nd) in which it was argued that *any* theory of descent would connect the various classes of facts equally well. To this Darwin replied in a characteristic letter. It was evident that he was most reluctant to continue the controversy, but thought it fair to admit publicly the force of his opponent's arguments.

In 1864 the Copley Medal of the Royal Society was given to Darwin. At the anniversary dinner of the Society, after the meeting at which the medals are presented by the President, Sir Charles Lyell in his speech made a "confession of faith" as to the "Origin." Darwin was prevented by illness from receiving the medal in person and from being present at the dinner.

The tenth edition of the "Principles" was published in 1867 and 1868, and in it Lyell clearly stated his belief in evolution. Sir Joseph Hooker, in his presidential address to the British Association at Norwich in 1868, eloquently spoke of the "new foundation" with which Lyell had under-pinned the edifice he had raised, and had thus rendered it "not only more secure, but more harmonious in its proportion than it was before." Wallace, too, in an article in the *Quarterly Review* (April, 1869), spoke with equal eloquence and force of the significance of Lyell's change of opinion.

Lyell's death took place in 1875, eleven years after his definite acceptance of Darwin's views. Darwin, in writing to Miss Arabella Buckley (now Mrs. Fisher, formerly secretary to Sir Charles Lyell), fully acknowledged the deep debt which he owed to

Lyell's teachings: "I never forget that almost everything which I have done in science I owe to the study of his great works." Huxley says in his obituary of Charles Darwin (reprinted in "Darwiniana," 1893, p. 268): "It is hardly too much to say that Darwin's greatest work is the outcome of the unflinching application to Biology of the leading idea and the method applied in the 'Principles' to Geology." Every biologist who realises—as who can help realising?—the boundless opportunities which Darwin's work has opened for him, will feel that he too owes a deep personal debt to Darwin's great teacher.

CHAPTER XVII.

INFLUENCE OF DARWIN UPON HOOKER AND ASA GRAY
—NATURAL SELECTION AND DESIGN IN NATURE
(1860–68).

HOOKER wrote on November 21st, speaking of the "glorious book" in the warmest terms. Later on in the year he wrote again in the same spirit, but speaking of the difficulty he found in assimilating the immense mass of details: "It is the very hardest book to read, to full profits, that I ever tried—it is so cram-full of matter and reasoning." Hooker must, however, have been familiar with the arguments and proofs, and for this reason did not attempt any detailed discussion. It is unnecessary to say more of Hooker's reception of the "Origin." During their long friendship Darwin had discussed the difficulties and the evidences of his theory more fully with him than with any other man; and, as "a man sharpeneth the countenance of his friend," the influence of Hooker was one of the most potent forces under which Darwin produced the greatest work of his life.

Many years later, when Hooker was awarded, in 1887, the Copley Medal of the Royal Society, reviewing his past experiences and work in his speech at the anniversary dinner, he concluded by telling us that his long and intimate friendship with Charles Darwin was the great event of his scientific career.

In sending a copy to Asa Gray, he wrote (November 11th):—

"I fully admit that there are very many difficulties not satisfactorily explained by my theory of descent with modification, but I cannot possibly believe that a false theory would explain so many classes of facts as I think it certainly does explain. On these grounds I drop my anchor, and believe that the difficulties will slowly disappear."

Asa Gray's reply was contained in a letter to Hooker, written January 5th, 1860, four days after reading the "Origin." He asks that Darwin may be told of what he had written. He says that the book "is done in a *masterly manner*. It might well have taken twenty years to produce it." He expressed the intention of reviewing the book, and seeing that Darwin and Hooker had fair play in America. A little later (January 23rd) he wrote to Darwin about the American reprint, etc., and spoke of the work itself in somewhat greater detail:—

"The *best part*, I think, is the *whole*, *i.e.* its *plan* and *treatment*, the vast amount of facts and acute inferences handled as if you had a perfect mastery of them. . . . Then your candour is worth everything to your cause. It is refreshing to find a person with a new theory who frankly confesses that he finds difficulties. . . . The moment I understood your premises, I felt sure you had a real foundation to hold on. . . . I am free to say that I never learnt so much from one book as I have from yours."

He considered that the attempt to account for the formation of organs such as eyes by natural selection, was the weakest point in the book. This view is to be explained by his strong teleological convictions.

Although Asa Gray was the great exponent of the " Origin " in America, he could not agree with Darwin on one important point—viz. on the exclusion of the ordinary conceptions of design in nature by the principle of natural selection. He believed that the two conceptions could be reconciled, and that design in some way worked in and through natural selection. By design is here meant what Huxley called "the commoner and coarser form of teleology," and which he believed to be now refuted—" the teleology which supposes that the eye, such as we see it in man or one of the higher vertebrata, was made with the precise structure it exhibits for the purpose of enabling the animal which possesses it to see, has undoubtedly received its death-blow." Huxley goes on to point out that there is a " wider teleology, which . . . is actually based upon the fundamental proposition of evolution . . . that the whole world . . . is the result of the mutual interaction, according to definite laws, of the forces possessed by the molecules of which the primitive nebulosity of the universe was composed." Therefore, " a sufficient intelligence could, from a knowledge of the properties of the molecules of that vapour, have predicted, say, the state of the fauna of Britain in 1869, with as much

H

certainty as one can say what will happen to the vapour of the breath on a cold winter's day." ("Genealogy of Animals," *The Academy*, 1869, reprinted in "Critiques and Addresses," and quoted in his chapter "On the Reception of the 'Origin of Species'" in the "Life and Letters," Vol. II.)

But at the time of the appearance of the "Origin," many who sympathised with the general drift of the argument were not yet prepared for the "wider teleology." Of these Asa Gray may be taken as the representative; and it will be of interest to follow the controversy between him and Darwin as regards design and natural selection. The recently published "Letters of Asa Gray to Charles Darwin" (Macmillan) enable us to follow the correspondence from the side of the great American evolutionist.

Writing November 26th, 1860, Darwin refers to one of Asa Gray's articles on the "Origin" :—

"I grieve to say that I cannot honestly go as far as you do about Design. I am conscious that I am in an utterly hopeless muddle. I cannot think that the world, as we see it, is the result of chance; and yet I cannot look at each separate thing as the result of Design. To take a crucial example, you lead me to infer—that you believe 'that variation has been led along certain beneficial lines.' I cannot believe this; and I think you would have to believe, that the tail of the Fantail was led to vary in the number and direction of its feathers in order to gratify the caprice of a few men. Yet if the Fantail had been a wild bird, and had used its abnormal tail for some special end, as to sail before the wind, unlike other birds, everyone would have said, 'What a beautiful and designed adaptation.' Again, I say I am, and shall ever remain, in a hopeless muddle."

Elsewhere Darwin suggested that the pouter pigeon, if it occurred wild, and used its inflated crop as a float, would be considered as a striking example of design.

This controversy between them continued for many years. We find Asa Gray referring to the argument of the pigeons three years later. Thus he wrote (September 1st, 1863):—

"I will consider about fantastic variation of pigeons. I see afar trouble enough ahead quoad design in nature, but have managed to keep off the chilliness by giving the knotty questions a rather wide berth. If I rather avoid, I cannot ignore the difficulties ahead. But if I adopt your view boldly, can you promise me any less difficulties?"

Writing the concluding paragraphs of the "Variations of Animals and Plants under Domestication," Darwin evidently bore in mind his controversies on the subject with Asa Gray and Lyell, and the attacks of the Duke of Argyll and others. Sending advanced sheets to Asa Gray, he wrote on October 16th, 1867:—

"I finish my book with a semi-theological paragraph, in which I quote and differ from you; what you will think of it, I know not."

In relation to this interesting controversy, I think it well to quote, almost in full, the metaphor by which Darwin enforced his argument that the origin of species by natural selection precluded a belief in design in nature as it was ordinarily conceived at the time.

This metaphor forms an important part of the conclusion of the work in question ("Variation of Animals and Plants," etc.) :

"The long-continued accumulation of beneficial variations will infallibly have led to structures as diversified, as beautifully adapted for various purposes and as excellently coordinated, as we see in the animals and plants around us. Hence I have spoken of selection as the paramount power, whether applied by man to the formation of domestic breeds, or by nature to the production of species. I may recur to the metaphor given in a former chapter : if an architect were to rear a noble and commodious edifice, without the use of cut stone, by selecting from the fragments at the base of a precipice wedged-formed stones for his arches, elongated stones for his lintels, and flat stones for his roof, we should admire his skill and regard him as the paramount power. Now, the fragments of stone, though indispensable to the architect, bear to the edifice built by him the same relation which the fluctuating variations of organic beings bear to the varied and admirable structures ultimately acquired by their modified descendants.

"Some authors have declared that natural selection explains nothing, unless the precise cause of each slight individual difference be made clear. If it were explained to a savage utterly ignorant of the art of building, how the edifice had been raised stone upon stone, and why wedge-formed fragments were used for the arches, flat stones for the roof, &c. ; and if the use of each part and of the whole building were pointed out, it would be unreasonable if he declared that nothing had been made clear to him, because the precise cause of the shape of each fragment could not be told. But this is a nearly parallel case with the objection that selection explains nothing, because we know not the cause of each individual difference in the structure of each being."

"The shape of the fragments of stone at the base of our precipice may be called accidental, but this is not strictly correct ; for the shape of each depends on a long sequence of events, all obeying natural laws. . . . But in regard to the use to which the fragments may be put, their shape may be strictly said to be accidental. . . ."

In his article in the *Nation* (March 19th, 1868), Asa Gray criticised the metaphor as follows :—

" But in Mr. Darwin's parallel, to meet the case of nature according to his own view of it, not only the fragments of rock (answering to variation) should fall, but the edifice (answering to natural selection) should rise, irrespective of will or choice ! "

This passage is quoted in the "Life and Letters" (Vol. III., p. 84), and Francis Darwin makes the convincing reply :—

" But my father's parallel demands that natural selection shall be the architect, not the edifice—the question of design only comes in with regard to the form of the building materials."

Darwin's reply was contained in his letter to Asa Gray dated May 8th, 1868 :—

" You give a good slap at my concluding metaphor : undoubtedly I ought to have brought in and contrasted natural and artificial selection ; but it seemed so obvious to me that natural selection depended on contingences even more complex than those which must have determined the shape of each fragment at the base of my precipice. What I wanted to show was that, in reference to pre-ordainment, whatever holds good in the formation of an English pouter-pigeon holds good in the formation of a natural species of pigeon. I cannot see that this is false. If the right variations occurred, and no others, natural selection would be superfluous."

To this, Asa Gray replied in his letter of May 25th :—

"As to close of my article, to match close of your book,—you see plainly I was put on the defence by your reference to an old hazardous remark of mine. I found your stone-house argument unanswerable in substance (for the notion of design must after all rest mostly on faith, and on accumulation of adaptations, &c.); so all I could do was to find a vulnerable spot in the shaping of it, fire my little shot, and run away in the smoke.

"Of course I understand your argument perfectly, and feel the might of it."

From this last letter I think we may conclude that Asa Gray's feelings on this subject rested, as he says, "on faith," and that, intellectually, he saw no way of meeting Darwin's arguments.

CHAPTER XVIII.

INFLUENCE OF DARWIN UPON HUXLEY.

It is of the utmost interest to trace the influence of Darwin upon Huxley, his great General in the numerous controversial battles which had to be fought before the new views were to secure a fair hearing and, at length, complete success. Now that we are quietly enjoying the fruit of his many victories, we are apt to forget how much we owe to Huxley, not only for evolution, but for that perfect freedom in the expression of thought and opinion which we enjoy. For Huxley fought on wider issues than those raised by evolution, wide as these are; and with a success so great that it is inconceivable that any new and equally illuminating thought which the future may hold in store for us, will meet with a reception like that accorded to the "Origin of Species."

At first sight it seems a simple matter to describe the effect of the "Origin" upon Huxley, considering that he, more than any other man, expounded it, and defended it from the most weighty of the attacks made upon it. Hence, it is only natural to believe, as many have done, that he was in entire agreement with the conclusions of the book as regards natural selection as well as evolution. On the other hand, the opinion has often been expressed that Huxley, although agreeing with the "Origin" for some years after its first appearance, changed his mind in later years, and no longer supported Darwin's views.

I shall give reasons for rejecting both these opinions about Huxley, although the first is far nearer the truth than the second. The latter is clearly untenable, and was probably merely an inference from the fact that after a time Huxley ceased to enter into Darwinian controversies. But this was because he had done his work with entire success, and therefore turned his attention in other directions. Whenever he was called on to write or speak about Darwinism, as he was on two occasions within a few months of his death, his writings and speeches left no doubt about his thoughts on the subject. Furthermore, in the Preface to "Darwiniana," written in 1893, he expressly denied that he had recanted or changed his opinions about Darwin's views.

In order to appreciate the influence of Darwin upon Huxley, we must find out the beliefs of the latter upon the "species question" before the appearance of the "Origin." In his chapter "On the Reception of the Origin of Species'" ("Life and Letters," Vol. II.) Huxley says that, before 1858, he took up an agnostic position as regards evolution ". . . upon two grounds: firstly, that up to that time, the evidence in favour of transmutation was wholly insufficient; and, secondly, that no suggestion respecting the causes of the transmutation assumed, which had been made, was in any way adequate to explain the phenomena." It is obvious that these two grounds are entirely distinct, and that the logical foundation of the first is far more secure than that of the second.

The effect of the "Origin" was completely to convince Huxley on the first ground: from that time he never doubted the truth of evolution, however it may have been brought about. With regard to the second ground, it is quite clear that Huxley had a very high opinion of natural selection: he thought it incomparably the best suggestion upon the subject that had ever been made, and he firmly believed that it accounted for something—that it may even have taken a dominant part in bringing about evolution. On the other hand, he never felt quite confident about the entire sufficiency of the evidence in its favour. It is probable that he was far more interested in the establishment of evolution as a fact than in natural selection as an explanation of it. He saw the vast amount of research in all kinds of new or almost neglected lines, which would be directly inspired by evolution. And his own investigations in some of these lines soon afforded some of the most weighty evidence in favour of the doctrine. Natural selection had not the same personal interest for him; no one has expounded it better or defended it more vigorously and successfully, but Huxley's own researches never lay in directions which would have made them available as a test of the theory. Of natural selection he might have used the words of Mercutio—it may not be "so deep as a well, nor so wide as a church door," to contain the whole explanation of evolution, "but 'tis enough 'twill serve"; it will, at any rate, prevent him from feeling the second ground on which he had maintained an agnostic position.

I believe that he maintained these views with inflexible consistency throughout his life, the only indications of change being in the last year, when the contrast between his certainty of evolution and his uncertainty of natural selection, as expressed in the two speeches quoted on pp. 140, 141, was, perhaps, more sharply marked than at any other period.

It is now proposed to support this conclusion by many extracts from Huxley's writings, as well as from his speeches, which have been alluded to above. The deep interest of the subject, and the wide differences of opinion with regard to it, justify, and indeed demand, copious quotations selected from works and speeches, written and spoken at many different times during the years between 1858 and 1894.

It may not be out of place to emphasise the fact that the sole responsibility for the conclusions here drawn rests with the author of this volume, and that the evidence on which the conclusions rest is supplied in full.

About a month before the "Origin" was published, Darwin wrote to Professor Huxley asking for the names of foreigners to whom to send his book. This communication is of great interest as being the earliest letter, accessible to the public, which he wrote to Huxley. In it he says: "I shall be *intensely* curious to hear what effect the book produces on you"; but he evidently thought that Huxley would disagree with much in it, and must have been surprised as well as gratified at the way in which it was received. In his chapter " On the Reception of the ' Origin of Species ' " (" Life and Letters," Vol. II.), Huxley writes: " My reflection, when I first made myself master of the central idea of the ' Origin, was, ' How extremely stupid not to have thought of that.' "

Huxley replied on November 23rd, 1859—the day before the publication of the " Origin "—saying that he had finished the book on the previous day. His letter was a complete acceptance of evolution as apart from any theory which may account for it; and a thorough agreement with natural selection as a "true cause for the production of species." At no time in his life did he state how far he considered natural selection to be a sufficient cause. He was only "prepared to go to the stake, if requisite, in support of" the chapters which marshal the evidence for evolution (ix., and most parts of x., xi., and xii).

With regard to the earlier chapters, which propound the theory of natural selection, his exact words are :—

"As to the first four chapters, I agree thoroughly and fully with all the principles laid down in them. I think you have demonstrated a true cause for the production of species, and have thrown the *onus probandi*, that species did not arise in the way you suppose, on your adversaries."

Darwin replied with much warmth, and expressed himself as "Now contented and able to sing my *Nunc Dimittis*."

In the *Times* of December 26th, 1859, appeared a masterly article upon the "Origin," and, after a time, it became known that Huxley was its author. Volume II. of the "Life and Letters" explains the circumstances under which the review was written. The article is reprinted as the first essay ("The Darwinian Hypothesis," I.) in "Darwiniana" (Vol. II. of the "Collected Essays of Professor Huxley," London, 1893). The following quotation (pp. 19, 20) shows the attitude he took up with regard to natural selection :—

"That this most ingenious hypothesis enables us to give a reason for many apparent anomalies in the distribution of living beings in time and space, and that it is not contradicted by the main phenomena of life and organisation appear to us to be unquestionable ; and, so far, it must be admitted to have an immense advantage over any of its predecessors. But it is quite another matter to affirm absolutely either the truth or falsehood of Mr. Darwin's views at the present stage of the enquiry. Goethe has an excellent aphorism defining that state of mind which he calls "Thätige Skepsis"---active doubt. It is doubt which so loves truth that it neither dares rest in doubting, nor extinguish itself by unjustified belief; and we commend this state of mind to students of species, with respect to Mr. Darwin's or any other hypothesis as to their origin. The combined investigations of another twenty years may, perhaps, enable naturalists to say whether the modifying causes and the selective power, which Mr. Darwin has satisfactorily shewn to exist in Nature, are competent to produce all the effects he ascribes to them ; or whether, on the other hand, he has been led to over-estimate the value of the principle of natural selection, as greatly as Lamarck over-estimated his *vera causa* of modification by exercise."

Of all the statements about natural selection made by Huxley, this one seems to me the nearest to the spirit of the two speeches he made in 1894, in which it became evident that the intervening thirty-five years had not brought the increased confidence he had hoped for. Furthermore, in the Preface to "Darwiniana" (1893) he expressly stated that he had not changed his mind as regards this article and the next which will be considered (see p. 137, where the passage is quoted).

In 1860 Huxley wrote the article on " The Origin of Species " which appeared in the *Westminster Review* for April, and is reprinted in "Darwiniana." He here states the reasons for his doubts about natural selection in considerable detail. At the beginning of the essay ("Darwiniana," p. 23) he asserts that—

" . . . all competent naturalists and physiologists, whatever their opinions as to the ultimate fate of the doctrines put forth, acknowledge that the work in which they are embodied is a solid contribution to knowledge and inaugurates a new epoch in natural history."

Towards the end of the essay, after vindicating the logical method followed by Darwin, he continues (pp. 73–75) :—

"There is no fault to be found with Mr. Darwin's method, then ; but it is another question whether he has fulfilled all the conditions imposed by that method. Is it satisfactorily proved, in fact, that species may be originated by selection ? that there is such a thing as natural selection ? that none of the phœnomena exhibited by species are inconsistent with the origin of species in this way ? If these questions can be answered in the affirmative, Mr. Darwin's view steps out of the ranks of hypotheses into those of proved theories ; but, so long as the evidence at present adduced falls short of enforcing that affirmation, so long, to our minds, must the new doctrine be content to remain among the former—an extremely valuable, and in the highest degree probable, doctrine, indeed the only extant hypothesis which is worth anything in a scientific point of view ; but still a hypothesis, and not yet the theory of species.

"After much consideration, and with assuredly no bias against Mr. Darwin's views, it is our clear conviction that, as the evidence stands, it is not absolutely proven that a group of animals, having all the characters exhibited by species in Nature, has ever been originated by selection, whether artificial or natural. Groups having the morphological character of species, distinct and permanent races in fact, have been so produced over and over again ; but there is no positive evidence, at present, that any group of animals has, by variation and selective breeding, given rise to another group which was even in the least degree infertile with the first. Mr. Darwin is perfectly aware of this weak point, and brings forward a multitude of ingenious and important arguments to diminish the force of the objection. We admit the value of these arguments to their fullest extent ; nay, we will go so far as to express our belief that experiments, conducted by a skilful physiologist, would very probably obtain the desired production of mutually more or less infertile breeds from a common stock, in a comparatively few years; but still, as the case stands at present, this 'little rift within the lute' is not to be disguised nor overlooked."

He concludes with a summary of the results of his argument. The sentences which bear on the present question are as follows (pp. 77, 78):—

"Our object has been attained if we have given an intelligible, however brief, account of the established facts connected with species, and of the relation of the explanation of those facts offered by Mr. Darwin to the theoretical views held by his predecessors and his contemporaries, and, above all, to the requirements of scientific logic. We have ventured to point out that it does not, as yet, satisfy all those requirements; but we do not hesitate to assert that it was superior to any preceding or contemporary hypothesis, in the extent of observational and experimental basis on which it rests, in its rigorously scientific method, and in its power of explaining biological phenomena, as was the hypothesis of Copernicus to the speculations of Ptolemy. But the planetary orbits turned out to be not quite circular after all, and, grand as was the service Copernicus rendered to science, Kepler and Newton had to come after him. What if the orbit of Darwinism should be a little too circular? what if species should offer residual phenomena, here and there, not explicable by natural selection? Twenty years hence naturalists may be in a position to say whether this is, or is not, the case; but in either event they will owe the author of 'The Origin of Species' an immense debt of gratitude. We should leave a very wrong impression on the reader's mind if we permitted him to suppose that the value of that work depends wholly on the ultimate justification of the theoretical views which it contains. On the contrary, if they were disproved to-morrow, the book would still be the best of its kind—the most compendious statements of well-sifted facts bearing on the doctrine of species that has ever appeared."

It is clear that two very distinct points are urged in this criticism of natural selection—(1) the difficulty that selective methods applied by man have not as yet produced all the characteristics of true species; (2) supposing the latter difficulty to be surmounted or sufficiently explained, the uncertainty as to how much or how little of the process of evolution has been due to natural selection.

Later in the same year Darwin seems to have been a little disappointed that Huxley's confidence did not increase. Thus, he wrote on December 2nd, 1860 :—

"I entirely agree with you that the difficulties on my notions are terrific ; yet having seen what all the *Reviews* have said against me, I have far more confidence in the *general* truth of the doctrine than I formerly had. Another thing gives me confidence—viz. that some who went half an inch with me now go further, and some who were bitterly opposed are now less bitterly opposed. And this makes me feel a little disappointed that you are not inclined to think the general view in some slight degree more probable than you did at first. This I consider rather ominous. Otherwise I should be more contented with your degree of belief. I can pretty plainly see that if my view is ever to be generally adopted, it will be by young men growing up and replacing the old workers, and then young ones finding that they can group facts and search out new lines of investigation better on the notion of descent than on that of creation."

In 1863 Huxley delivered a course of lectures to working men on "The Causes of the Phenomena of Organic Nature"; here, too, he expressed his opinions about natural selection with great clearness and force. These lectures are reprinted as the concluding part of "Darwiniana," and the references are to the pages of that volume of his collected essays.

On page 464 we read—

"Here are the phenomena of Hybridism staring you in the face, and you cannot say, 'I can, by selective modification, produce these same results.' Now, it is admitted on all hands, at present, so far as experiments have gone, it has not been found possible to produce this complete physiological divergence by selective breeding. . . . If we were shewn that this must be the necessary and inevitable results of all experiments, I hold that Mr. Darwin's hypothesis would be utterly shattered."

He then goes on to show that this is very far from proved, and concludes (page 466)—

"that though Mr. Darwin's hypothesis does not completely extricate us from this difficulty at present, we have not the least right to say it will not do so."

A passage on page 467 shows that Huxley placed natural selection infinitely higher than any other attempt to account for evolution, and indeed that he regarded all other attempts with scorn.

"I really believe that the alternative is either Darwinism or nothing, for I do not know of any rational conception or theory of the Organic universe which has any scientific position at all beside Mr. Darwin's. . . . Whatever may be the objections to his views, certainly all other theories are absolutely out of court."

I

On page 468 he continues—

"But you must recollect that when I say I think it is either Mr. Darwin's hypothesis or nothing; that either we must take his view, or look upon the whole of organic nature as an enigma, the meaning of which is wholly hidden from us; you must understand that I mean that I accept it provisionally, in exactly the same way as I accept any other hypothesis."

He concludes the lectures and the volume in which they are now reproduced by the following eloquent testimony to the unique value of the "Origin of Species":—

"I believe that if you strip it of its theoretical part it still remains one of the greatest encyclopædias of biological doctrine that any one man ever brought forth, and I believe that, if you take it as the embodiment of an hypothesis, it is destined to be the guide of biological and psychological speculation for the next three or four generations."

The next essay from which I quote was written in 1871. At the beginning of "Mr. Darwin's Critics" ("Darwiniana," p. 120) he uses words which, if they stood alone, might be interpreted as an indication of a stronger conviction.

"Whatever may be thought or said about Mr. Darwin's doctrines, or the manner in which he has propounded them, this much is certain, that, in a dozen years, the 'Origin of Species' has worked as complete a revolution in biological science as the 'Principia' did in astronomy—and it has done so, because, in the words of Helmholtz, it contains an 'essentially new creative thought.'"

This last quotation, and the following one, from "Evolution in Biology," written in 1878, are, I think, among the strongest utterances in favour of natural selection to be found in the Collected Essays. At the conclusion of the above-named essay (*l. c.*, p. 223) he states that it was clearly seen that—

"if the explanation would apply to species, it would not only solve the problem of their evolution, but that it would account for the facts of teleology, as well as for those of morphology ; . . . "

"How far 'natural selection' suffices for the production of species remains to be seen. Few can doubt that, if not the whole cause, it is a very important factor in that operation ; and that it must play a great part in the sorting out of varieties into those which are transitory and those which are permanent."

The seventh essay, "The Coming of Age of 'The Origin of Species,'" was written in 1880. His complete confidence in evolution, as shown in this essay, may be contrasted with his cautious statements about natural selection. He boldly affirms evolution to be the fundamental doctrine of the "Origin of Species," while natural selection is, I believe, neither mentioned nor even alluded to. On this great occasion he thus emphasised the immense debt we owe to Darwin in that he was the first to produce adequate evidence in favour of the ancient doctrine of evolution, a benefit quite distinct from that which he conferred in the theory of natural selection (see pp. 100–102).

The following are among the most confident statements about evolution to be found in this essay. Speaking of the "Origin," he says (p. 229) :—

" . . . the general doctrine of evolution, to one side of which it gives expression, obtains, in the phenomena of biology, a firm base of operations whence it may conduct its conquest of the whole realm of nature."

And again, on page 332 :—

"The fundamental doctrine of the 'Origin of Species,' as of all forms of the theory of evolution applied to biology, is 'that the innumerable species, genera, and families of organic beings with which the world is peopled have all descended, each within its own class or group, from common parents, and have all been modified in the course of descent.'"

Furthermore, on page 242 we read :—

"I venture to repeat what I have said before, that so far as the animal world is concerned, evolution is no longer a speculation, but a statement of historical fact. It takes its place alongside of those accepted truths which must be reckoned with by philosophers of all schools."

And on the same page he quotes with approval the statement by M. Filhol of the results to which he had been led by his palæontological investigations :—

"Under the influence of natural conditions of which we have no exact knowledge, though traces of them are discoverable, species have been modified in a thousand ways : species have arisen which, becoming fixed, have thus produced a corresponding number of secondary species."

Similarly, in the Obituary notice in *Nature* (1882), Huxley speaks of the secure position in which Darwin had placed the doctrine of evolution as his great achievement. The following eloquent passage occurs on page 247 of "Darwiniana" :—

"None have fought better, and none have been more fortunate, than Charles Darwin. He found a great truth trodden underfoot, reviled by bigots, and ridiculed by all the world ; he lived long enough to see it, chiefly by his own efforts, irrefragibly established in science, . . . "

In the impressive speech in which Huxley handed over the statue of Darwin to the Prince of Wales, as representative of the Trustees of the British Museum, on June 9th, 1885 ("Darwiniana," p. 248), the references to Darwin are most consistent with the view that the support to evolution was held by the speaker to be the great work of his life. Natural selection is not mentioned.

The next publication on this subject by Huxley is the celebrated chapter "On the Reception of the 'Origin of Species,'" in the second volume of the great "Life and Letters." In this chapter he speaks rather more confidently about natural selection than in some of the earlier essays and in the later speeches:—

"The reality and the importance of the natural processes on which Darwin founds his deductions are no more doubted than those of growth and multiplication ; and, whether the full potency attributed to them is admitted or not, no one doubts their vast and far-reaching significance."

But of evolution he speaks far more strongly:—

"To any one who studies the signs of the times, the emergence of the philosophy of Evolution, ["bound hand and foot and cast into utter darkness during the millennium of theological scholasticism "] in the attitude of claimant to the throne of the world of thought, from the limbo of hated and, as many hoped, forgotten things, is the most portentous event of the nineteenth century."

And for this he gives Darwin the credit.

Later on he indicates the sense in which his keen appreciation of natural selection is to be understood. Thus, such strong statements as—

" . . . the publication of the Darwin and Wallace papers in 1858, and still more that of the 'Origin' in 1859, had the effect . . . of the flash of light, which to a man who has lost himself in a dark night, suddenly reveals a road which, whether it takes him straight home or not, certainly goes his way";

and—

"The facts of variability, of the struggle for existence, of adaptation to conditions, were notorious enough; but none of us had suspected that the road to the heart of the species problem lay through them, until Darwin and Wallace dispelled the darkness, and the beacon-fire of the 'Origin' guided the benighted,"

if they stood alone, might naturally be interpreted as an unqualified testimony to the permanent truth of natural selection. But this interpretation is expressly excluded :—

" Whether the particular shape which the doctrine of evolution, as applied to the organic world, took in Darwin's hands, would prove to be final or not, was, to me, a matter of indifference. In my earliest criticisms of the 'Origin' I ventured to point that its logical foundation was insecure . . . ; and that insecurity remains."

Its value for Huxley was that it was "incomparably more probable than the creation hypothesis"; that it was "a hypothesis respecting the origin of known organic forms, which assumed the operation of no causes but such as could be proved to be actually at work"; that it provided "clear and definite conceptions which could be brought face to face with facts and have their validity tested"; that it freed us "for ever from the dilemma—refuse to accept the creation hypothesis, and what have you to propose that can be accepted by any cautious reasoner?" Indeed, the hypothesis did away with this dilemma, even if it were itself to disappear; for "if we had none of us been able to discern the paramount significance of some of the most patent and notorious of natural facts, until they were, so to speak, thrust under our noses, what force remained in the dilemma —creation or nothing? It was obvious that, hereafter, the probability would be immensely greater, that the links of natural causation were hidden from our purblind eyes, than that natural causation should be incompetent to produce all the phenomena of nature."

Therefore, "the only rational course for those who had no other object than the attainment of truth, was to accept 'Darwinism' as a working hypothesis, and see what could be made of it." Furthermore, " Whatever may be the ultimate fate of the particular theory put forth by Darwin, . . . all the ingenuity and all the learning of hostile critics has not enabled them to adduce a solitary fact, of which it can be said, this is irreconcilable with the Darwinian theory."

Taking this argument as a whole, it seems to me to amount to the words of Mercutio quoted at the beginning of this chapter.

In the following year (1888) Huxley wrote the Obituary Notice of Darwin for the Proceedings of the Royal Society : it is reprinted in "Darwiniana" (pp. 253 *et seq.*). In this admirable essay the author recognises that Darwin evidently accepted evolution before he could offer any explanation of the motive cause by which that process took place. The theory of descent with modification had often been thought of before, " but in the eyes of the naturalist of the ' Beagle ' (and, probably, in those of most sober thinkers), the advocates of transmutation had done the doctrine they expounded more harm than good." Huxley speaks of the " Origin " as " one of the hardest books to master," in this agreeing with Hooker (see p. 111).

In this essay Huxley gives a clear and excellent statement of natural selection, prefaced by these words (p. 287) :—

"Although, then, the present occasion is not suitable for any detailed criticism of the theory, or of the objections which have been brought against it, it may not be out of place to endeavour to separate the substance of the theory from its accidents ; and to shew that a variety not only of hostile comments, but of friendly would-be improvements, lose their *raison d'être* to the careful student."

Then follows a brief but epigrammatic description, such as only Huxley could have written, of the theory, and some of the chief arguments which have revolved round it. Occasionally he speaks as if he were stating his own opinion as well as Darwin's, but throughout it seems to me that his object is not to give his own views but to write a fair and clear account of Darwin's theory, and to defend it from a number of criticisms and modifications which have been, from time to time, brought forward.

"Darwiniana" was published in 1893, and this is the date of the Preface, in which Huxley speaks of—

" . . . the ancient doctrine of Evolution, rehabilitated and placed upon a sound scientific foundation, since, and in consequence of, the publication of the 'Origin of Species . . .'"

He thinks that readers will admit that in the first two essays (see pages 124–128 of the present volume)—

" . . . my zeal to secure fair play for Mr. Darwin, did not drive me into the position of a mere advocate ; and that, while doing justice to the greatness of the argument, I did not fail to indicate its weak points. I have never seen any reason for departing from the position which I took up in these two essays ; and the assertion which I sometimes meet with nowadays, that I have 'recanted' or changed my opinions about Mr. Darwin's views, is quite unintelligible to me."

"As I have said in the seventh essay, [see pages 131, 132 of the present volume] the fact of evolution is to my mind sufficiently evidenced by palæontology ; and I remain of the opinion expressed in the second, that until selective breeding is definitely proved to give rise to varieties infertile with one another, the logical foundation of the theory of natural selection is incomplete."

It is therefore clear, as I have before stated, that Huxley, in 1893, re-stated his criticisms and qualifications of thirty years before, and expressed his conviction anew of the validity of the objections which he then raised against a full and complete acceptance of natural selection.

We now come to the last and most significant of all Huxley's utterances on evolution and natural selection, made on two great occasions in the last year of his life. Lord Salisbury, in his eloquent and interesting Presidential Address to the British Association at Oxford (August 8th, 1894), had said of Darwin :—

"He has, as a matter of fact, disposed of the doctrine of the immutability of species. . . . Few now are found to doubt that animals separated by differences far exceeding those that distinguish what we know as species have yet descended from common ancestors."

While thus completely admitting evolution in the organic world, Lord Salisbury attacked natural selection on two grounds—first, on the insufficiency of the time allowed by physicists for a process which is, of necessity, extremely slow in its operation; secondly, on the ground that "we cannot demonstrate the process of natural selection in detail; we cannot even, with more or less ease, imagine it." And his main objection under this head was the supposed difficulty in securing the union of successful variations. The actual words have been already quoted on page 83, where it was shown that the criticism does not apply to natural selection, but to a theory mistaken by the speaker for that of Darwin. Curiously enough, the first objection of the insufficiency of time was the indirect cause of a subsequent trenchant criticism by Professor Perry of the line of mathematical reasoning on which the limit had been fixed.

Huxley was called on to second the vote of thanks, and his speech had evidently been considered with the greatest care. I quote the passages which bear on evolution and natural selection from the *Times* of August 9th, 1894, in which a *verbatim* report is furnished :—

" . . . As one of those persons who for many years past had made a pretty free use of the comfortable word ' evolution,' let him remind them that 34 years ago a considerable discussion, to which the President had referred, took place in one of their sectional meetings upon what people frequently called the ' Darwinism question,' but which on that occasion was not the Darwinism question, but the very much deeper question which lay beneath the Darwinism question — he meant the question of evolution. . . . The two doctrines, the two main points, for which these men [Sir John Lubbock, Sir J. Hooker, and the speaker] fought were that species were mutable, and that the great variety of animal forms had proceeded from gradual and natural modification of the comparatively few primitive forms. . . . ."

After alluding to the revolution in thought which had taken place in thirty-four years, he said :—

"As he noted in the presidential address to which they had just listened with such well-deserved interest, he found it stated on that which was then and at this time the highest authority for them, that as a matter of fact the doctrine of the immutability of species was disposed of and gone. He found that few were now found to doubt that animals separated by differences far exceeding those which they knew as species were yet descended from a common ancestry. Those were their propositions ; those were the fundamental principles of the doctrine of evolution. Darwinism was not evolution, nor Spencerism, nor Hæckelism, nor Weismannism, but all these were built on the fundamental doctrine which was evolution, which they maintained so many years, and which was that upon which their President had put the seal of his authority that evening. . . . ."

Huxley thus hailed the statements of the President in favour of evolution, while the attacks on natural selection he merely met by saying that the address would have made a good subject for discussion in one of the sections, and by insisting with impressive solemnity that evolution was a very different thing from natural selection, thereby implying that the former would be unaffected by the fate of the latter.

The second occasion was between three and four months later, when Huxley spoke at the Anniversary Dinner of the Royal Society, November 30th, 1894, after having been awarded the Darwin Medal at the afternoon meeting. I quote his words from the *verbatim* report of the *Times* for December 1st :—

" . . . I am as much convinced now as I was 34 years ago that the theory propounded by Mr. Darwin, I mean that which he propounded—not that which has been reported to be his by too many ill-instructed, both friends and foes— has never yet been shewn to be inconsistent with any positive observations, and if I may use a phrase which I know has been objected to and which I use in a totally different sense from that in which it was first proposed by its first propounder, I do believe that on all grounds of pure science it 'holds the field,' as the only hypothesis at present before us which has a sound scientific foundation. . . . I am sincerely of opinion that the views which were propounded by Mr. Darwin 34 years ago may be understood hereafter as constituting an epoch in the intellectual history of the human race. They will modify the whole system of our thought and opinion, our most intimate convictions. But I do not know, I do not think anybody knows, whether the particular views which he held will be hereafter fortified by the experience of the ages which come after us ; . . . whether the particular form in which he has put them before us (the Darwinian doctrines) may be such as is finally destined to survive or not is more, I venture to think, than anybody is capable at this present moment of saying."

It is unnecessary to say anything about this passage, which fitly sums up and sets the seal on the long series of quotations I have felt obliged to make.

It may not be out of place, however, to state in a few words why many naturalists, including the present writer, are not inclined to accept the extremely cautious and guarded language of one upon whom, with regard to so many other subjects, they have ever looked as their teacher and guide. Concerning the verification of a hypothesis, Huxley said in his lectures to working men ("Darwiniana," pages 367, 368)—

" . . . that the more extensive verifications are,—that the more frequently experiments have been made, and results of the same kind arrived at,—that the more varied the conditions under which the same results are attained, the more certain is the ultimate conclusion . . . ."

And again—

"In scientific enquiry it becomes a matter of duty to expose a supposed law to every possible kind of verification, and to take care, moreover, that this is done intentionally, and not left to a mere accident . . . ."

It may well be that the length of time required before an artificially-selected race will exhibit, when interbred with the parent species, phenomena of hybridism similar to those which are witnessed when distinct natural species are interbred—will be fatal to the production of this important line of evidence. But there is nothing to hinder us from holding the reasonable belief that such evidence might be obtained if we had command of the necessary conditions; and in the meantime other evidence of the most satisfactory kind is accumulating, and on a vast scale. Whenever a naturalist approaches a problem in the light of the theory of natural selection, and is able, by its aid, to predict a conclusion which subsequent investigation proves to be correct, he is helping in the production of evidence in favour of the theory. When a naturalist has found the formula "if natural selection be true so-and-so ought to happen" the safest of all guides into the unknown, when it has brought him success many times and in very different directions, when he knows that many other workers in other fields of biological inquiry have had a similarly happy experience, he gradually comes to feel a profound confidence in the permanent truth and the far-reaching importance of the great theory which has served him so well.

CHAPTER XIX.

EVEN earlier than Huxley, H. C. Watson wrote warmly accepting natural selection. In his letter, which is dated November 21st, 1859, he said :—

"Your leading idea will surely become recognised as an established truth in science—*i.e.* 'Natural Selection.' It has the characteristics of all great natural truths, clarifying what was obscure, simplifying what was intricate, adding greatly to previous knowledge. You are the greatest revolutionist in natural history of this century, if not of all centuries."

For some years to come, however, such views as these were the exception, as will soon be shown.

The Duke of Argyll has argued (*Nineteenth Century*, December, 1887) that the success of "Natural Selection" has followed from the convincing character of the words used, scientific men ("the populace of science" he calls them) being so easily led by the power of loose analogies that they have been convinced of the truth of the principle because they are familiar with Nature on the one hand, and selection as a process on the other!

As I am not aware that this preposterous suggestion has ever been publicly disproved, and since therefore some readers of the journal in question may have been misled by it, I have collected much evidence, which proves that the principle of natural selection was only absorbed with the very greatest difficulty, and that the words used in describing it for a long time entirely failed to inform even eminent scientific men of the essential characteristics of the theory itself, and certainly failed most signally to convince them. Conviction came very gradually as the theory was slowly understood and was seen to offer an intelligible explanation of an immense and ever-increasing number of facts.

I will now bring together quotations from Darwin's letters in 1859 and 1860, showing how soon he came to realise the difficulty with which natural selection was understood, and to feel that he might have been more successful with some other title.

In 1859 he wrote to Dr. W. B. Carpenter—" I have found the most extraordinary difficulty in making even able men understand at what I was driving." The remaining quotations are all taken from letters written in 1860. By the middle of this year, when he was feeling oppressed by hostile reviews and unfair and ignorant criticisms (" I am getting wearied at the storm of hostile reviews, and hardly any useful "), he often alludes to the failure of opponents to understand his theory. Thus, in a letter to Hooker (June 5th), he says :—

" This review, however, and Harvey's letter have convinced me that I must be a very bad explainer. Neither really

J

understand what I mean by Natural Selection. . . . . I hope to God you will be more successful than I have been in making people understand your meaning."

He says almost the same thing in a letter to Lyell (June 6th) :—

" . . . . I am beginning to despair of ever making the majority understand my notions. . . . . I must be a very bad explainer. I hope to Heaven that you will succeed better. Several reviews and several letters have shown me too clearly how little I am understood. I suppose 'Natural Selection' was a bad term ; . . . . I can only hope by reiterated explanations finally to make the matter clearer."

Writing to Asa Gray, he says :—

" . . . . I have had a letter of fourteen folio pages from Harvey against my book, with some ingenious and new remarks ; but it is an extraordinary fact that he does not understand at all what I mean by Natural Selection."

Later on, he again wrote to Lyell :—

" Talking of 'natural selection' ; if I had to commence *de novo*, I would have used 'natural preservation.' For I find men like Harvey of Dublin cannot understand me, though he has read the book twice. Dr. Gray of the British Museum remarked to me that, ' *selection* was obviously impossible with plants ! No one could tell him how it could be possible ! ' And he may now add that the author did not attempt it to him ! "

And still later he wrote asking Lyell's advice as to additions to a new edition of the " Origin," saying :— " I would also put a note to ' Natural Selection,' and show how variously it has been misunderstood." This note is to be found on page 63 of the sixth edition. In it he tells us that some writers have "even imagined that natural selection induces variability," instead of merely preserving it; others that natural selection "implies conscious choice in the animals which become modified"; others that it is set up "as an active power or Deity." In writing (December) to Murray about a new edition of the "Origin," he alludes to the "many corrections, or rather additions, which I have made in hopes of making my many rather stupid reviewers at least understand what is meant."

He seems to have retained a very vivid recollection of the difficulty with which his theory was understood at first; thus he tells us in his "Autobiography":—

"I tried once or twice to explain to able men what I meant by Natural Selection, but signally failed."

Why the term "natural selection" was chosen by Darwin is very clearly shown in the three following quotations from letters to distinguished scientific men, which were probably written in answer to attacks or criticisms on this very point.

He writes to Lyell in 1859, "Why I like the term is that it is constantly used in all works on breeding."

Writing to H. G. Bronn in 1860, he explains his motives with great clearness and force :—

"Several scientific men have thought the term 'Natural Selection' good, because its meaning is *not* obvious, and each man could not put on it his own interpretation, and because it at once connects variation under domestication and nature. . . . Man has altered, and thus improved the English race-horse by *selecting* successive fleeter individuals ; and I believe, owing to the struggle for existence, that similar *slight* variations in a wild horse, *if advantageous to it*, would be *selected* or *preserved* by nature ; hence Natural Selection."

In 1866 he wrote to Wallace, comparing the term with that which we owe to Herbert Spencer :—

" I fully agree with all that you say on the advantages of H. Spencer's excellent expression of ' the survival of the fittest.' This however had not occurred to me till reading your letter. It is, however, a great objection to this term that it cannot be used as a substantive governing a verb ; and that it is a real objection I infer from H. Spencer continually using the words, natural selection.   I formerly thought, probably in an exaggerated degree, that it was a great advantage to bring into connection natural and artificial selection ; this indeed led me to use a term in common, and I still think it some advantage. . . . The term Natural Selection has now been so largely used abroad and at home, that I doubt whether it could be given up, and with all its faults I should be sorry to see the attempt made.   Whether it will be rejected must now depend ' on the survival of the fittest.' As in time the term must grow intelligible the objections to its use will grow weaker and weaker.   I doubt whether the use of any term would have made the subject intelligible to some minds, clear as it is to others ; for do we not see even to the present day Malthus on Population absurdly misunderstood ?   This reflection about Malthus has often comforted me when I have been vexed at the mis-statement of my own views."

A large number of critics not only failed to understand natural selection, but they asserted that it was precisely the same theory as that advanced by Lamarck or one of the other writers on evolution before Darwin.   This seems almost incredible to us at the present day, when the biological world is divided into two sections on the very subject, and when it is generally recognised that Lamarck's theory would be, if it were proved to be sound, a formidable rival to natural selection as a motive cause of evolution.   But the following quotations—a few among many—leave no doubt whatever upon the subject.

Evidence on this point reached Darwin almost immediately after the appearance of the "Origin." Thus he writes to Hooker on December 14th, 1859 :—

"Old J. E. Gray, at the British Museum, attacked me in fine style : 'You have just reproduced Lamarck's doctrine, and nothing else, and here Lyell and others have been attacking him for twenty years, and because *you* . . . say the very same thing, they are all coming round ; it is the most ridiculous inconsistency,' &c. &c."

In the following year, Wilberforce, Bishop of Oxford, writing in the *Quarterly Review* for July, 1860, appeals to Lyell,

"in order that with his help this flimsy speculation may be as completely put down as was what in spite of all denials we must venture to call its twin though less-instructed brother, the 'Vestiges of Creation.'"

Again, Dr. Bree, in "Species not Transmutable," says :

"The only real difference between Mr. Darwin and his two predecessors, [Lamarck and the "Vestiges"] is this :—that while the latter have each given a mode by which they conceive the great changes they believe in have been brought about, Mr. Darwin does no such thing."

One of the most interesting of the countless examples of misunderstanding is contained in a recently published letter from W. S. Macleay to Robert Lowe.* This letter was written from Elizabeth Bay, and is dated May, 1860, evidently just after the first edition of the "Origin," a copy of which had been sent by Robert Lowe, had been read by Macleay.

"Again if this primordial cell had a Creator, as Darwin seems to admit, I do not see what we gain by denying the Creator, as Darwin does, all management of it after its creation. Lamarck was more logical in supposing it to have existed of itself from all eternity—indeed this is the principal difference that I see between this theory of Darwin's and that of Lamarck, who propounded everything essential in the former theory, in a work now rather rare—his 'Philosophie Zoologique.' But you may see an abridgment of it in so common a book as his 'Histoire Nat. des Animaux Vertébrés,' vol. i., pp. 188, *et seq.*—Edit. 1818, where the examples given of natural selection are the gasteropod molluscs. . . . Natural selection (sometimes called 'struggles' by Darwin) is identical with the 'Besoins des Choses' of Lamarck, who, by means of his hypothesis, for instance, assigns the constant stretching of the neck to reach the acacia leaves as the cause of the extreme length of it in the giraffe; much in the same way the black bear, according to Darwin, became a whale, which I believe as little as his other assertion that our progenitors anciently had gills—only they had dropped off by want of use in the course of myriads of generations."

I had long been anxious to possess a copy of the first edition of the "Origin," and was fortunate enough to come across one about the time when Macleay's letter was pointed out to me by my wife.

* "Life of Lord Sherbrooke," Vol. II. (pp. 205-206), Longmans & Co. London, 1893.

I opened the title-page, and found upon it the signature " W. S. Macleay"; it must have been the very volume given him by Robert Lowe, which Macleay had read and believed he had been fairly criticising. Out of Macleay's volume, therefore, I quote the sentences he referred to in his letter.

Darwin's real statement about the black bear which "became a whale" is to be found on page 184 :—

"In North America the black bear was seen by Hearne swimming for hours with widely open mouth, thus catching, like a whale, insects in the water. Even in so extreme a case as this, if the supply of insects were constant, and if better adapted competitors did not already exist in the country, I can see no difficulty in a race of bears being rendered, by natural selection, more and more aquatic in their structure and habits, with larger and larger mouths, till a creature was produced as monstrous as a whale."

The statement about the gills which "dropped off by want of use" becomes in the original (p. 191):—

"In the higher vertebrata the branchiæ have wholly disappeared—the slits on the sides of the neck and the loop-like course of the arteries still marking in the embryo their former position."

Although the hypothetical case of the black bear—carefully guarded as it is—does not now appear to us at all extravagant (indeed, in the cleft cheeks of the goat-sucker we have a precisely analogous case), Darwin seems to have thought it unsuitable, probably because it became an easy butt for ignorant ridicule. We find accordingly that in the second and all subsequent editions everything after the word " water " is omitted, while "almost" is inserted before "like a whale." He was alluding to this passage when he wrote to Lyell (December 22nd, 1859): "Thanks about 'Bears,' a word of ill-omen to me." Furthermore, Andrew Murray * says, concerning the sentences as they stand in the first edition:—

"In quoting this, I do not at all mean to give it as a fair illustration of Mr. Darwin's views. I only refer to it as indicating the extent to which he is prepared to go. The example here given I look upon (as I have reason to know that Mr. Darwin himself does) merely as an extreme and somewhat extravagant illustration, imagined expressly to show in a forcible way how 'natural selection' would operate in making a mouth bigger and bigger, because more advantageous."

* Proc. Roy. Soc. Edin., Jan. 16th, 1860.

CHAPTER XX.

THE history of opinion on evolution and natural
selection, in the years which followed the publication
of the " Origin," can be traced in the titles of the
papers and subjects of discussion at successive
meetings of the British Association. In the Presidential Address delivered by Professor Newton to
the Biological Section of the Manchester meeting in
1887, there is a most interesting account of the
struggles which took place :—

" The ever-memorable meeting . . . at Oxford in the summer
of 1860 saw the first open conflict between the professors of
the new faith and the adherents of the old one. Far be it
from me to blame those among the latter who honestly stuck
to the creed in which they had educated themselves ; but my
admiration is for the few dauntless men who, without flinch-
ing from the unpopularity of their cause, flung themselves in
the way of obloquy, and impetuously assaulted the ancient
citadel in which the sanctity of ' species ' was enshrined
and worshipped as a palladium. However strongly I myself
sympathised with them, I cannot fairly state that the conflict
on this occasion was otherwise than a drawn battle ; and thus
matters stood when in the following year the Association met
in this city [Manchester]. That, as I have already said, was a
time of ' slack water.' But though the ancient beliefs were

not much troubled, it was for the last time that they could be said to prevail ; and thus I look upon our meeting in Manchester 1861 as a crisis in the history of biology. All the same, the ancient beliefs were not allowed to pass wholly unchallenged ; and one thing is especially to be marked—they were challenged by one who was no naturalist at all, by one who was a severe thinker no less than an active worker ; one who was generally right in his logic, and never wrong in his instinct ; one who, though a politician, was invariably an honest man—I mean the late Professor Fawcett. On this occasion he brought the clearness of his mental vision to bear upon Mr. Darwin's theory, with the result that Mr. Darwin's method of investigation was shewn to be strictly in accordance with the rules of deductive philosophy, and to throw light where all was dark before."

Professor Newton specially alluded to this interesting case of Professor Fawcett as illustrating his conviction that the theory of natural selection—

"did not, except in one small point, require a naturalist to think it out and establish its truth. . . . But in order to see the effect of this principle upon organic life the knowledge— the peculiar knowledge—of the naturalist was required. This was the knowledge of those slight variations which are found in all groups of animals and plants. . . . Herein lay the triumph of Mr. Darwin and Mr. Wallace. That triumph, however, was not celebrated at Manchester. The question was of such magnitude as to need another year's incubation, and the crucial struggle came a twelvemonth later when the Association met at Cambridge. The victory of the new doctrine was then declared in a way that none could doubt. I have no inclination to join in the pursuit of the fugitives."

There is reason to believe that Professor Newton's impressions of the result of the celebrated meeting of the British Association at Oxford in 1860 are more accurate than those of the eyewitness quoted in the "Life and Letters." The latter has pictured a brilliant triumph for Huxley in the renowned duel with the Bishop of Oxford. But I have been told by more than one of the audience that Huxley was really too angry to speak effectively, nor is this to be wondered at, considering the extreme provocation. Mr. William Sidgwick, who was present and sympathised warmly with Huxley, has told me that this was his opinion. I have heard the same from the Rev. W. Tuckwell, who also quoted a remark of the late Professor Rolleston tending in the same direction. Mr. Tuckwell said that it was clear that the audience as a whole was not carried away by Huxley's speech, but, on the contrary, was obviously shocked at it; and he contrasted that occasion with another at which he was also present, in the North, several years later, when Huxley replied to an opponent who, like the bishop, appealed to the theological prejudices of his hearers. But by that time the new teachings had been absorbed, and Huxley gained a signal triumph.

It must not be supposed that Darwin was by any means indifferent to the attacks on his views. On the contrary, his sensitive nature was greatly depressed by the violent and often most unfair criticisms to which he was subjected, although beneath this evident disturbance lay the firm conviction that he had seen the truth, and that the truth would in the end be seen by others.

After the great fight with the bishop at the

British Association at Oxford, he wrote to Hooker (July 2nd, 1860) :—

"I have read lately so many hostile views, that I was beginning to think that perhaps I was wholly in the wrong, and that —— was right when he said the whole subject would be forgotten in ten years; but now that I hear that you and Huxley will fight publicly (which I am sure I never could do), I fully believe that our cause will, in the long-run, prevail."

Looking at the history of opinion on this subject, the slowness with which the new ideas were absorbed appears remarkable. Even so able a man as the late Professor Rolleston wrote in 1870 (" Forms of Animal Life," Introduction, p. xxv., First Edition) the following carefully guarded sentences, which, it is to be noted, deal with evolution rather than natural selection. Speaking of "the theory of evolution with which Mr. Darwin's name is connected," Rolleston says :—

"Many of the peculiarities which attach to biological classifications would thus receive a reasonable explanation ; but where verification is, *ex hypothesi*, impossible, such a theory cannot be held to be advanced out of the region of probability. The acceptance or rejection of the general theory will depend, as does the acceptance or rejection of other views supported merely by probable evidence, upon the particular constitution of each individual mind to which it is presented !"

It was too much to expect that many of the older scientific men would retain sufficient intellectual flexibility to be able to recognise, as Lyell had, that the facts of nature were explained and predicted better by the new views than by those in which they had grown up. Darwin thoroughly understood this, and, writing to his friends, maintained that the fate of his views was in the hands of the younger men.

A grand yet simple conception like that of natural selection, explaining and connecting together innumerable facts which people had previously explained differently, or had become accustomed to regard as inexplicable, must always remain as a stumbling-block to the majority of those who have reached or passed middle life before its first appearance.

Hardly anything is more characteristic of Darwin than the tone with which he wrote to acknowledged opponents. Thus his letters to L. Agassiz (1868), Quatrefages (1869 or 1870), and Fabre (1880), are models of the way in which a correspondence which would present peculiar difficulties to most people may be conducted. In these letters there is not the least attempt to slur over or minimise the points of wide difference; on the contrary, they are most candidly stated, but with so much respect and sympathy, and with such marked appreciation of the knowledge he had gained from his correspondent, that the reader must have regretted the divergence of opinion as greatly as the writer.

Tyndall has given a very interesting and pathetic account of the evident distress with which Professor L. Agassiz, chief of the opponents of Darwin in

America, recognised the success of the teachings he could not accept.

"Sprung from a race of theologians, this celebrated man combated to the last the theory of natural selection. One of the many times I had the pleasure of meeting him in the United States was at Mr. Winthrop's beautiful residence at Brookline, near Boston. Rising from luncheon, we all halted, as if by a common impulse in front of a window, and continued there a discussion which had been started at table. The maple was in its autumn glory; and the exquisite beauty of the scene outside seemed, in my case, to interpenetrate without disturbance the intellectual action. Earnestly, almost sadly, Agassiz turned and said to the gentlemen standing round, 'I confess that I was not prepared to see this theory received as it has been by the best intellects of our time. Its success is greater than I could have thought possible.'" *

The history of science can hardly supply anything more sad than the blight which may fall on a man's career because he is unable, from conscientious motives, to use some great means of advance. Such a weapon for the progress of science was provided by the Darwinian theory, and men were to be henceforth divided according to their use or neglect of the new opportunities. Men who up to that time had been equals were to be for ever separated, some to press forward in the front rank of scientific discovery, others to remain as interesting relics of a byegone age.

It is hardly necessary to say that this does not apply to men, like Agassiz, who had already left their mark deep upon the science of their day, but it has a

* Presidential address to the British Association at Belfast, 1874. Report, p. lxxxvii.

very real application to those men whose position was to be estimated by work done after the year 1858.

In the midst of those years of struggle and anxiety which followed the appearance of the "Origin," we meet with another instance of the same extraordinary foresight which appeared in his contention in favour of the persistence of the great oceans and continental areas. I refer to his views on spontaneous generation —a very ancient belief, and one which from time to time has been the will-o'-the-wisp of biological speculation, leading it into all kinds of fruitless and dangerous regions.*

Dr. Carpenter's "Introduction to the Study of Foraminifera" had been reviewed in the *Athenæum* (March 28th, 1863), the writer attacking evolution and favouring spontaneous generation, or, as it was then called, heterogeny. Darwin wrote to Hooker, who had lent him a copy of the paper, " Who would have ever thought of the old stupid *Athenæum* taking to Oken-like transcendental philosophy written in Owenian style ! . . . It will be some time before we see ' slime, protoplasm, etc.,' generating a new animal. . . . It is mere rubbish, thinking at present of the origin of life ; one might as well think of the origin of matter." In 1871 he wrote :—

"It is often said that all the conditions for the first production of a living organism are now present, which could ever have been present. But if (and oh ! what a big if !) we could conceive in some warm little pond, with all sorts of

* See H. F. Osborn, " From the Greeks to Darwin " (1894).

ammonia and phosphoric salts, light, heat, electricity, etc., present, that a proteine compound was chemically formed ready to undergo still more complex changes, at the present day such matter would be instantly devoured or absorbed, which would not have been the case before living creatures were formed."

About 1870 Dr. H. C. Bastian began working on the subject, and brought forward in his "Origin of of Lowest Organisms" (1871), and "The Beginnings of Life" (1872), what he believed to be conclusive evidence of the truth of spontaneous generation, for which he proposed the term Archebiosis. His enthusiasm and strong convictions were contagious, and for a time the belief spread rather widely, although it soon collapsed before the researches and arguments of Pasteur, Tyndall, and Huxley. Darwin read "The Beginnings of Life," and wrote about it to Wallace (August 28th, 1872) as follows:—

"His [Bastian's] general argument in favour of Archebiosis is wonderfully strong, though I cannot think much of some few of his arguments. The result is that I am bewildered and astonished by his statements, but am not convinced, though, on the whole, it seems to me probable that Archebiosis is true. I am not convinced, partly I think owing to the deductive cast of much of his reasoning ; and I know not why, but I never feel convinced by deduction, even in the case of H. Spencer's writings. . . . I must have more evidence that germs, or the minutest fragments of the lowest forms, are always killed by 212° of Fahr. . . . As for Rotifers and Tardigrades being spontaneously generated, my mind can no more digest such statements, whether true or false, than my stomach can digest a lump of lead."

CHAPTER XXI.

WE now come to consider the succession of invaluable works produced by Darwin after the appearance of the " Origin," the last of which—that on Earthworms— was published about six months before his death.

Darwin's method of bringing these results before the world was somewhat different from that most generally adopted by scientific men in this country, although of common occurrence in Germany. The great majority of scientific facts are here published by the proceedings or transactions of scientific societies, or in special journals; and although a scientific man frequently brings together his general results into a volume for the public, the original communications remain as the detailed exposition of his researches.

Darwin, too, wrote a very large number of memoirs for the scientific societies, as may be seen from the list in Appendix III. of the " Life and Letters," but the volumes which he subsequently published included *all* the previous details, with the addition of much new matter, and it is these volumes rather than the original communications which form the authoritative statement of his investigations. Such a method was

K

possible and desirable with the subjects upon which he worked, all of which were of great interest to the thinking part of the general public, as well as to the experts; but in less attractive subjects it is not probable that the plan could be carried out in this country with any prospect of success.

It has already been stated that Darwin looked on the " Origin of Species " as a short abstract of a greater work he intended to publish. It is likely that he at first contemplated a comprehensive work like the " Origin " itself, but soon found that his notes on domesticated animals and plants, the general results of which had been condensed into the first three chapters of the " Origin," would form a work more than twice the size of the latter. He began arranging these notes on January 9th, 1860 (January 1st is the date given in the " Autobiography "), as soon as the second edition of the " Origin " was off his hands, but his " enormous correspondence," as he calls it in the " Autobiography," with friends about the " Origin," and the reviews and discussions upon it, must have occupied a large part of his time ; and then there was the third edition to bring out (published April, 1861). This edition must have cost much labour, as many parts were modified and enlarged to meet the objections or misunderstanding of reviewers.

Francis Darwin tells us that the third chapter of " Animals and Plants, &c., " was still on hand at the beginning of 1861. His work on this book was furthermore interrupted by illnesses and by other researches.

Thus, during 1860 he worked at Drosera, and during the latter part of 1861 and beginning of 1862 at the fertilisation of orchids. In his diary for 1866 we meet with the entry, " *Nov.* 21*st*—Finished ' Pangenesis,' " and later on, " *Dec.* 22*nd*—Began concluding chapter of book." In this year, too, he brought out the fourth edition of the " Origin." When the time for publication approached Darwin was much disappointed at the dimensions of the work. It was not published till January 30th, 1868, when it was proved that his fears were groundless, for a second edition of 1,250 copies were required in the following month, the 1,500 of the first edition having been all absorbed.

This work is considered by some writers to be the greatest produced by Darwin ; but I think we shall be right in accepting his own opinion that such words should be applied to the " Origin." It is probable, however, that this book stands second in importance in the splendid list of works which have done so much to increase our knowledge of nature and to inspire others to continue the good work.

" The Variation of Animals and Plants under Domestication " opens with a very clearly written account of natural selection ; it proceeds to treat of the domestic quadrupeds and birds, describing the differences between the various breeds of each species, and making out as far as possible the history of their development from each other and from the wild stock. Cultivated plants are then treated in the same manner. The first volume concludes with two most important chapters on bud-variation and anomalous modes of reproduction, and on inheritance.

The second volume deals with inheritance, crossing, effect of conditions of life, sterility, hybridism, selection by man, causes and laws of variability. Finally, all the main lines are brought to a common centre in the wonderful chapter in which he discloses his "provisional hypothesis of pangenesis." This is of such interest, and is so characteristic of its author's power of viewing the most divergent facts from a common standpoint, that it is desirable to give a tolerably full account of it.

The following is a brief statement of the various classes of facts which Darwin attempted to connect by his hypothesis.

Reproduction is sexual and asexual, and the latter is of various kinds, although their differences are more apparent than real. It may be concluded that gemmation or budding, fission or division, the repair of injuries, the maintenance of each part, and the growth of the embryo "are all essentially the results of one and the same great power."

In parthenogenesis the ovum can develop without fertilisation, and hence the union of germs from different individuals cannot serve as an essential characteristic of sexual, as compared with asexual, generation. Although sexually-produced individuals tend to vary far more than those which are produced asexually, this is not always the case, and the variability, when it occurs, is subject to the same laws.

Sexually-produced individuals very generally pass in development from a lower to a higher grade; but this can hardly be said to occur in certain forms, such as Aphis, etc.

The differences between the two forms of reproduction being thus much less than at first sight appears, we are led to inquire for the reason why the more complex and difficult process is so universal. Sexual reproduction appears to confer two benefits on organisms—(1) " When species are rendered highly variable by changed conditions of life, the free intercrossing of the varying individuals will tend to keep each form fitted for its proper place in nature, and crossing can be effected only by sexual generation "; (2) Many experiments tend to show that free and wide inter-crossing induces vigour in the offspring.

Darwin concludes that the reason why the germ-cell perishes if it does not unite with another from the opposite sex is simply because it includes " too little formative matter for independent existence and development." He was led to this conclusion by the fact that the male and female germ-cells " do not in ordinary cases differ in their power of giving character to the embryo," and also from experiments which seemed to show that a certain number of pollen grains or of spermatozoa may be required to fertilise a single seed or ovum. " The belief that it is the function of the spermatozoa to communicate life to the ovule seems a strange one, seeing that the

unimpregnated ovule is already alive, and continues
for a considerable time alive."

It is very remarkable to note how largely Professor Weismann's conclusions on this subject were
anticipated by this part of Darwin's work.

Graft hybrids.—The probability that a graft may
alter the character of the stock to which it is united,
so that hybrid buds might be formed by budding or
grafting the tissues of distinct varieties or species,
would, if it became a certainty, prove the essential
identity of sexual and asexual reproduction; " for the
power of combining in the offspring the characters
of both parents is the most striking of all the
functions of sexual generation."

Direct action of the male element on the female.—
Pollen from another species is known to affect the
mother-plant in certain cases. Thus pollen from the
lemon has caused stripes of lemon-peel in the fruit
of the orange; the peel is, of course, formed by
the mother-plant, and is quite different from
the part which the male element is adapted to
affect—viz. the ovule. Similar cases are known
among animals, as in the celebrated example of
Lord Morton's mare.

Development.—The changes by which the embryo
reaches maturity differ immensely, even within the
limits of the same compact group. Forms which
closely resemble each other in the mature state, and
are intimately related to each other, such as the various species of lobster and crayfish, etc.,.pass through

a totally different developmental history. Hence we are led to believe in the complete independence of "each structure from that which precedes and follows it in the course of development."

The functional independence of the elements or units of the body. Variability and inheritance.— Variability generally results "from changed conditions acting during successive generations." The influence is exerted on the sexual system, and if extreme, impotence tends to be produced. Bud-variation proves that "variability is not necessarily connected with the sexual system." The inherited effects of use and disuse of parts imply that the changes in the cells of a distant part of the body affect the reproductive cells, so that the being produced from one of these cells inherits the changes. "Nothing in the whole circuit of physiology is more wonderful."

"Inheritance is the rule and non-inheritance the anomaly." Inheritance follows laws, such as the tendency for a character to appear at corresponding ages in parent and offspring. Reversion "proves to us that the transmission of a character and its development . . . . are distinct powers." Crossing strongly induces reversion. "Every character which occasionally reappears is present in a latent form in each generation."

The hypothesis of pangenesis attempts to explain and connect together all the facts and conclusions which have been summarised in the preceding pages.

This hypothesis assumes that each one of the countless cells of which the body of a higher animal is composed throws off a minute gemmule which, with those derived from other cells, exists in the body, and when supplied with nutriment multiplies by division. Each gemmule is capable of ultimate development into a cell similar to the one from which it, either directly or indirectly, arose. Each cell of the body dispatches its representative, as it were, to each single germ-cell, and this explains how it is that the latter possess the power of reproducing the likeness of the parent body. But the germ-cells also receive dormant gemmules which may remain undeveloped until some generation in the remote future. The development of the gemmules into cells depends on their union with the developing cells which precede them in the order of growth. Gemmules are thrown off during each stage of growth and during maturity.

This hypothesis of pangenesis is so called because the whole body is supposed to produce the elements from which new individuals arise, the germ-cells being only the union of these elements into clusters.

The fact that hybrids may be produced by grafting, that the pollen can act on the tissues of the female plant, and the male germ-cells on the future offspring of the female, implies that the reproductive material can exist and the reproductive processes take place in the tissues, and that they are not confined to the germ-cells.

The retention of dormant gemmules, and their passage from generation to generation until their development, may seem improbable; but is it more so than the *fact* which their presence would explain —viz. the transmission of latent structures and their ultimate reappearance?

The development of the whole plant from a Begonia leaf implies that these gemmules are very widely distributed through the tissues.

The elective affinity of the gemmules for the cells which precede them in growth may be paralleled by the affinity of the male and female germ-cells, as we see in the preference of a plant for the pollen grains of its own over those of closely-allied species, or by the attraction of the minute germs of disease to certain tissues of the body.

It is possible that the numerous gemmules thrown off by the cells of a complex structure, such as a feather, "may be aggregated into a compound gem-mule." In the case of a petal, however, where parts as well as the whole are apt to develop, as is seen in the case of "stripes of the calyx assuming the colour and texture of the corolla," it is more probable that the gemmules are separate and free. The cell itself is a complex structure, and we do not know whether its separate parts are not developed from the separate gemmules of an aggregate.

Such an hypothesis explains the fundamental similarity which has already been shown to exist between all modes of reproduction. The gemmules collected in bud or germ-cell are essentially similar; and were it not for the special advantages of sexual reproduction (increased vigour and more marked variation of offspring), we can well believe that it would have been much less general. The formation of graft-hybrids, and the action of the male element on the mother and on future offspring, become intelligible. The antagonism between growth and sexual reproduction in animals, and between increase by buds, etc., and seeds in plants, can be understood by the use of gemmules in one direction preventing their simultaneous use in another.

The regrowth of an amputated part, as in the salamander or snail, is explained by the presence and development of gemmules previously thrown off from the part. The difficulty that a limb is produced of the same age as that which was lost, and not a larval limb, and that the cells with which the gemmules must unite at first are not those which precede them in the course of growth, but mature cells, is met by the consideration that this power is a special one adapted to meet special dangers to certain parts of certain animals, and that it is therefore probable that appropriate provision has been made by natural selection: it may be in the form of "a stock of nascent cells or of partially developed gemmules." The existence of these latter in buds, and their absence from sexual cells, may account for bud development being the more direct and brief of the two. The much greater tendency to repair lost parts

in lower and younger forms may be due to the same
cause.

The occasional tendency of hybrids to resemble
one parent in one part and the other in another
may be due to superabundance of gemmules in the
fertilised germ, those from one parent having "some
advantage in number, affinity, or vigour over those
derived from the other parent." The general pre-
ponderance of one parent over the other may be
similarly explained. The cases in which " the colour
or other characters of either parent tend to appear
in stripes or blotches " are to be understood by the
gemmules having an affinity for others of the same
kind.

The sterility of hybrids is entirely due to the
reproductive organs being affected; in the case of
plants they continue to propagate freely by buds. The
hybrid cells throw off hybrid gemmules which collect
in the buds but cannot do so in the reproductive
organs.

Development and metamorphosis.—The remarkable facts of development and metamorphosis are well
explained by the hypothesis. Allied forms may pass
to a similar end through very dissimilar stages or
conversely. Parts may appear to develop within
previously existing corresponding parts, or they may
appear within parts which are quite distinct. These
divergent facts are explained by the hypothesis,
each part during each stage being formed independently from the gemmules of the same part

in previous generations, and not, although it may appear to do so, from the corresponding parts of earlier stages. In the process of time certain parts during certain stages may be affected by use or disuse or surroundings, and the parts of subsequent generations will be similarly affected, because formed from correspondingly altered gemmules; but this need not affect the other stages of the same parts.

Transposition and multiplication of parts.—The cases of abnormal transposition or multiplication of organs—for instance, the development of teeth in the palate or of pollen in the edge of a petal—are to be explained by supposing that the gemmules unite with wrong cells instead of, or as well as, the right ones; " and this would follow from a slight change in their elective affinities." Such slight changes are known to occur; for instance, certain plants " absolutely refuse to be fertilised by their own pollen, though abundantly fertile with that of any other individual of the same species." Inasmuch as the cells of adjoining parts will often have nearly the same structure, we can understand that some slight change in elective affinity may affect a large area. Hence we can account for a crowd of horns on the head of a sheep, or many spurs on the leg of a fowl, etc. Frequently repeated parts are extremely liable to vary in number; in this case we have a large number of closely allied gemmules and of points for their union, and slight changes in elective affinity would be specially apt to occur.

VARIABILITY.—Changed conditions may lead to irregularity in the number of gemmules derived from various parts of the body; deficiency in number might cause variation in any part by leaving some of the cells free to unite with allied gemmules.

The direct action of surroundings, or the effect of use or disuse on a part, may cause corresponding modifications of the gemmules, and through these of the part in the succeeding generation. " A more perplexing problem can hardly be proposed," and yet it receives an explanation on this hypothesis. Such causes must, as a rule, act during many generations before the modification reappears in the offspring. This may be due to the unaltered gemmules derived from earlier generations, and their gradual replacement by the increasing number of altered gemmules.

Variation in plants is much more frequent in sexually produced than it is in asexually produced individuals. This may be due to the absence in the latter of that great cause of variability, changes in the reproductive organs under altered conditions. Furthermore, the former alone pass through the earlier phases of development, when structure is most plastic and yields most readily to the causes inducing variability.

The stability of hybrids and of many varieties when propagated by buds, as compared with their reversion to the parent form when propagated by seed, remains inexplicable.

Hence variability is explained as due (1) to the

irregularity in number of gemmules, to their trans-
positions, and redevelopment when dormant; and (2)
to their actual modification and the gradual replacement by them of unaltered gemmules.

INHERITANCE.—The non-transmission by heredity
of mutilations, even when repeated for many generations, as in docking the tails of certain domesticated
breeds, may be explained by the persistence of gemmules from still earlier generations. The cases of
inheritance when mutilations are followed by disease,
as in Brown-Séquard's experiments on guinea-pigs,
may be due to the gemmules being attracted to the
diseased part and there destroyed.

The disappearance of a rudimentary part, together
with its occasional reappearance by reversion, is to be
understood by the existence of ancestral gemmules,
for which the corresponding cells have, except in the
cases of reversion, lost their affinity. When the
disappearance is final and complete, the gemmules
have probably perished altogether.

"Most, or perhaps all, of the secondary characters
which appertain to one sex, lie dormant in the other
sex; that is, gemmules capable of development into
the secondary male sexual characters are included
within the female; and conversely female characters
in the male." This is seen in cases of castration or
when the sexual organs from any cause have become
functionless. The sex in which such changes are
brought about tends to develop the secondary sexual
characters of the other sex. The normal develop-

ment of the secondary characters proper to the sex
of the individual may be explained by a slight
difference in the elective affinity of the cells so that
they attract the corresponding gemmules rather than
those of the opposite sex, which as we have seen
are also present.

The male characters of the male sex are in many
species latent except at certain seasons of the year,
and in both sexes the proper characters are latent
until sexual maturity. All such latent characters are
closely connected with the cases of ordinary reversion.
The appearance (whether seasonal or in the course of
development) of cells with affinities for the latent
gemmules explains the development of the characters
in question.

Certain butterflies and plants (*e.g.* Lythrum) produce two or more separate forms of individuals. In
these cases each individual includes the latent gemmules of the other forms as well as its own. Her-
maphroditism in unisexual species, and especially in
the occasional cases of insects in which the right side
of the body is one sex and the left side the other, the
line of separation dividing the individual into two
equal halves, can be explained by slight abnormal
changes in the affinities of cells for gemmules, so
that a certain group of cells, or all the cells on one
side of the body, attract the gemmules which would
normally have remained latent.

Reversion is induced by a change of conditions
and especially by crossing. The first results of crossing

are usually intermediate between the parents, but in the next generation there is commonly reversion to one or both parent-forms, or even to a more remote ancestor. The existence of abundant hybridised gemmules is shown by the propagation of the cross in a true form by means of buds; but dormant gemmules from the parent-form are also present and multiply. In the sexual elements of the hybrid there are both pure and hybrid gemmules, and the addition of the pure gemmules in one sex to those in the other accounts for the reversion, especially if we assume that pure "gemmules of the same nature would be especially apt to combine." Partial reversion on the one hand, and the reappearance of the hybrid form on the other, would be respectively due to a combination of pure with hybrid gemmules, and of the hybrid gemmules from both parent hybrids.

When characters which do not blend exist in the parents, crossing may result in an insufficiency of gemmules from the male alone and from the female alone, and then dormant ancestral gemmules might have the opportunity of development, and thus cause reversion. Similarly certain conditions might favour the increase and development of dormant gemmules. Diseases appearing in alternate generations, or gaining strength by the intermission of a generation, may be due to the increase of the gemmules in the intervening time, and the same explanation may hold for the sudden and irregular increase of a weakly inherited modification.

Darwin ends his general conclusions with these words :—

"No other attempt, as far as I am aware, has been made, imperfect as this confessedly is, to connect under one point of view these several grand classes of facts. An organic being is a microcosm—a little universe, formed of a host of self-propagating organisms, inconceivably minute and numerous as the stars in heaven."

CHAPTER XXII.

PANGENESIS AND CONTINUITY OF THE GERM-PLASM :
DARWIN'S CONFIDENCE IN PANGENESIS.

DARWIN's letters prove that he thought very highly
of this hypothesis; and whether the future determine
it to be true or erroneous, it must surely rank as
among the greatest of his intellectual efforts. In his
autobiography he says of it :—

"An unverified hypothesis is of little or no value; but if
any one should hereafter be led to make observations by which
some such hypothesis could be established, I shall have done
good service, as an astonishing number of isolated facts can be
thus connected together and rendered intelligible."

The hypothesis was submitted to Huxley (May
27th, 1865 ?) in manuscript and alluded to in the
letter sent at the same time. An unfavourable reply
was evidently received, for we find Darwin writing to
Huxley, July 12th (1865 ?) :—

" I do not doubt your judgment is perfectly just, and I will
try to persuade myself not to publish. The whole affair is
much too speculative ; yet I think some such view will have
to be adopted, when I call to mind such facts as the inherited
effects of use and disuse, &c."

This last sentence is of great interest, and the
same opinion comes out strongly in his published

account of the hypothesis, viz. the view that the real facts which imperatively demand some material to pass from the body-cells to the germ-cells in order to account for their hereditary transmission are the effects of use and disuse, or the influence of surroundings—in fact, all those characters which are now called "acquired." And it is impossible to escape the conclusion that, if acquired characters are transmissible by heredity, an hypothesis which is substantially that of pangenesis will have to be accepted. Darwin did not doubt this transmission, and he framed pangenesis mainly to account for it.

Considerable doubt has of recent years been thrown upon the transmission of acquired characters, and if hereafter this doubt be justified, it will be possible to substitute for pangenesis a hypothesis like the "continuity of the germ-plasm" brought forward by Professor Weismann. A few words indicating the contrast between the two hypotheses may not be out of place.

In Professor Weismann's hypothesis the germ-plasm contained in the nucleus of the germ-cell possesses, if placed under right conditions, the power of developing into an organism. It is not, however, entirely used up during development, and the part which remains grows and is stored in the germ-cells of the offspring, and ultimately develops into the succeeding generation. Hence parent and offspring resemble each other because they are formed from the same

thing.   There is no real break between the generations; they are thrown up successively from a continuous line of germ-plasm.   In this hypothesis the germ
is the essential thing, the body a mere secondary pro-
duct.   It is a theory of Blastogenesis as contrasted
with Pangenesis.   The hereditary transmission of acquired characters, in which many still believe, is quite
irreconcilable with it, and if substantiated would over-
throw it altogether.

On the other hand the body-cells are the essential
elements of pangenesis, and the germ-cells the mere
meeting-places of their representatives and quite
devoid of significance on their own account.   There
is some sort of interruption between successive
generations, as the gemmules develop into cells,
which again throw off gemmules; the break, however, is bridged by the ancestral gemmules and by
the life of the body-cell which intervenes between
the gemmule from which it arose and that to which
it gives rise.

The remaining chief occasions on which Darwin
alludes to pangenesis in his published letters are
quoted below; they prove his confidence in the
hypothesis and the nature of the hold it had upon his
mind.

Later on he again wrote to Huxley on the same
subject :—

"I am rather ashamed of the whole affair, but not converted
to a no-belief. . . . It is all rubbish to speculate as I have
done; yet, if I ever have strength to publish my next book, I

fear I shall not resist ' Pangenesis,' but I assure you I will put
it humbly enough. The ordinary course of development of
beings, such as the Echinodermata, in which new organs are
formed at quite remote spots from the analogous previous parts,
seems to me extremely difficult to reconcile on any view except
the free diffusion in the parent of the germs or gemmules of
each separate new organ : and so in cases of alternate generation."

To LYELL, *August 22nd*, 1867.

" I have been particularly pleased that you have noticed
Pangenesis. I do not know whether you ever had the feeling
of having thought so much over a subject that you had lost all
power of judging it. This is my case with Pangenesis (which is
26 or 27 years old), but I am inclined to think that if it be
admitted as a probable hypothesis it will be a somewhat
important step in Biology."

To ASA GRAY, *October 16th*, 1867.

" The chapter on what I call Pangenesis will be called a
mad dream, and I-shall be pretty well satisfied if you think it
a dream worth publishing ; but at the bottom of my own mind
I think it contains a great truth."

To HOOKER, *November 17th* [1867].

" I shall be intensely anxious to hear what you think about
Pangenesis ; though I can see how fearfully imperfect, even in
mere conjectural conclusions, it is ; yet it has been an infinite
satisfaction to me somehow to connect the various large groups
of facts, which I have long considered, by an intelligible
thread."

To FRITZ MÜLLER, *January 30th* [1868].

" . . . . I should very much like to hear what you think of
' Pangenesis,' though I fear it will appear to *every one* far too
speculative."

To HOOKER, *February* 23*rd* [1868].

After expressing a fear that Pangenesis is still-born because of the difficulty with which it is understood, he says :—

"You will think me very self-sufficient, when I declare that I feel *sure* if Pangenesis is now still-born it will, thank God, at some future time reappear, begotten by some other father, and christened by some other name. Have you ever met with any tangible and clear view of what takes place in generation, whether by seeds or buds, or how a long-lost character can possibly reappear; or how the male element can possibly affect the mother plant, or the mother animal, so that her future progeny are affected? Now all these points and many others are connected together, whether truly or falsely is another question, by Pangenesis. You see I die hard, and stick up for my poor child."

To WALLACE, *February* 27*th* [1868].

"You cannot well imagine how much I have been pleased by what you say about ' Pangenesis.' . . . What you say exactly and fully expresses my feeling, viz. that it is a relief to have some feasible explanation of the various facts, which can be given up as soon as any better hypothesis is found. It has certainly been an immense relief to my mind ; for I have been stumbling over the subject for years, dimly seeing that some relation existed between the various classes of facts. . . . You have indeed pleased me, for I had given up the great god Pan as a still-born deity."

To HOOKER, *February* 28*th* [1868].

"I see clearly that any satisfaction which Pan may give will depend on the constitution of each man's mind. . . . I heard yesterday from Wallace, who says (excuse horrid vanity), 'I can hardly tell you how much I admire the chapter on "Pangenesis." It is a *positive comfort* to me to have any feasible explanation of a difficulty that has always been haunting me, and I shall never be able to give it up till a better one supplies its place, and that I think hardly possible, &c.' Now his foregoing [italicised] words express my sentiments exactly and fully: though perhaps I feel the relief extra strongly from having during many years vainly attempted to form some hypothesis. When you or Huxley say that a single cell of a plant, or stump of an amputated limb, has the ' potentiality' of reproducing the whole—or ' diffuses an influence,' these words give me no positive idea ;—but, when it is said that the cells of a plant, or stump, include atoms derived from every other cell of the whole organism and capable of development, I gain a distinct idea. But this idea would not be worth a rush, if it applied to one case alone ; but it seems to me to apply to all the forms of reproduction—inheritance—metamorphosis—to the abnormal transposition of organs—to the direct action of the male element on the mother plant, &c. Therefore I fully believe that each cell does *actually* throw off an atom or gemmule of its contents ;—but whether or not, this hypothesis serves as a useful connecting link for various grand classes of physiological facts, which at present stand absolutely isolated."

To V. CARUS, *March* 21st [1868].

". . . . Sir C. Lyell says to every one, 'You may not believe in "Pangenesis," but if you once understand it, you will never get it out of your mind.' And with this criticism I am perfectly content. All cases of inheritance and reversion and development now appear to me under a new light."

To FRITZ MÜLLER, *June*, 1868.

"I have yet hopes that you will think well of Pangenesis. I feel sure that our minds are somewhat alike, and I find it a great relief to have some definite, though hypothetical view, when I reflect on the wonderful transformations of animals,—the re-growth of parts,—and especially the direct action of pollen on the mother-form, &c. It often appears to me almost certain that the characters of the parents are 'photographed' on the child, only by means of material atoms derived from each cell in both parents, and developed in the child."

To Asa Gray, *May* 8*th* [1868].

"Your article in the *Nation* [March 19th] seems to me very good, and you give an excellent idea of Pangenesis—an infant cherished by few as yet, except his tender parent, but which will live a long life. There is parental presumption for you!"

To E. Ray Lankester, *March* 15*th* [1870].

"I was pleased to see you refer* to my much despised child, 'Pangenesis,' who I think will some day, under some better nurse, turn out a fine stripling."

To Wallace, *August* 28*th*, 1872.

"Notwithstanding all his [Dr. Bastian's] sneers I do not strike my colours as yet about Pangenesis."

In the second edition of "Animals and Plants," Beale's criticism of the hypothesis is alluded to with characteristic candour and humour :—"Dr. Lionel Beale (*Nature*, May 11th, 1871, p. 26) sneers at the whole doctrine with much acerbity and some justice." Galton's paper before the Royal Society (March 30th, 1871), upon the results of inter-transfusion of blood as destructive of pangenesis, was answered by Darwin in *Nature* (April 27th, 1871). He did "not allow that pangenesis has as yet received its death-blow, though from presenting so many vulnerable points its life is always in jeopardy."

Galton had argued that the gemmules present in the blood of one individual would be expected to pass into the other individual and to produce hereditary

* In " Comparative Longevity."

effects on its offspring. This, however, did not occur.
Romanes repeated these experiments under more
rigid conditions, but with the same negative results;
equally negative were the effects of his transplantation
of skin from one animal to another, although the
skin grew quite successfully in its new position.

CHAPTER XXIII.

THE work on " The Descent of Man " was begun as
soon as Darwin had sent the manuscript of " Animals
and Plants " to the printers, although notes on the
subject had been collected from time to time during
the previous thirty years—in fact, ever since Darwin
had come to definite conclusions on evolution.

The book was published February 24th, 1871, but,
as in the case of his other publications, continuous
work upon the manuscript had been impossible. The
volume attracted great interest, and 5,000 copies were
printed in 1871 in addition to the first 2,500.

The full title of the book is "The Descent of
Man, and Selection in Relation to Sex," and, as this
title almost implies, it is made up of two distinct
works, which might well have been issued separately.
The first part, dealing with man, is far shorter than
the other. Darwin had from the first considered
that his theory of evolution by natural selection
would involve man as well as the other animals, and,
that no one might accuse him of want of candour, he
had said in the " Origin" that by this work " light
would be thrown on the origin of man and his

history." But he was anxious to justify this statement, which was, of course, distasteful to very many in those days, by a most complete treatment of the subject.

He opens this part of the work, which he calls "The Descent or Origin of Man," by discussing the structures which are common to man and animals, including those which are represented in man in a rudimentary state, and by showing the similarity of the phases through which man and animals pass during their embryological development.

Having thus shown that man was probably descended from some lower form, he considers the mode by which the process was effected, showing that man possesses variability in body and mind, and is, like other animals, subject to all the laws of inheritance and variation, and to the direct action of surrounding conditions, and to the effect of the use and disuse of parts, and that his rate of increase is such as to render a large amount of extermination inevitable. In other words, he presents the same facilities for the operation of natural selection as those presented by other animals. The points in which man differs from other animals are then considered in relation to their possible origin by natural selection. The differences and resemblances between the mind of man and animals are discussed in much detail, and the origin of the former through natural selection is defended. This part concludes with the consideration of the position of man in the animal series, his birthplace and antiquity, and with an account of the formation of races.

In the second part Darwin brings forward a large body of evidence in favour of his hypothesis of sexual selection—viz. the view that, in the higher animals, some alteration, especially of the secondary sexual characters, is produced by the preferences and rejections of the sex which is sought by the other. Such results are commonly found in the males as a result of the preferences of the females accumulated through countless generations ; but in some species the females court the males, and are themselves subject to the same process of improvement by selection.

Opinion is still divided on this most interesting question. Wallace, more convinced than ever as to the efficiency and scope of natural selection, after first doubting, has finally come to reject sexual selection altogether. Probably the majority of naturalists are convinced by Darwin's arguments and his great array of facts that the principle of sexual selection is real, and accounts for certain relatively unimportant features in the higher animals, and they further accept Darwin's opinion that its action has always been entirely subordinate to natural selection.

A brief third part considers sexual selection in relation to man.

Darwin says, in his " Autobiography," that sexual selection and " the variation of our domestic productions, together with the causes and laws of variation, inheritance, and the intercrossing of plants, are the sole subjects which I have been able to write about in full, so as to use all the materials which I have collected."

"The Expression of the Emotions," at first intended as a chapter of the "Descent," was begun, only two days after the proofs of the latter had been corrected, on January 15th, 1871. The book was published in the autumn of the following year; the edition consisted of 7,000 copies, and 2,000 were printed at the end of the year; and this, we are told, was a mistake, as it prevented the appearance of a second edition, with notes and corrections, during the author's lifetime. Darwin had begun to take notes on this subject when his first child was born, December 27th, 1839, for he tells us that, even then, he felt convinced " that the most complex and fine shades of expression must all have had a gradual and natural origin."—(" Autobiography.")

In this work Darwin argues with great wealth of illustration and the record of numberless interesting observations, that the movements of expression are to be explained by three principles. The first of these is that movements made in gratifying some desire become by repetition so habitual that the slightest feeling of desire leads to their performance, however useless they may then be. The second principle is that of antithesis—" the habit of voluntarily performing opposite movements under opposite impulses."

The third principle is "the direct action of the
excited nervous system on the body, independently
of the will, and independently, in large part, of
habit."

By showing that the expressions of emotions could
thus be explained naturally, Darwin undermined the
position taken up by Sir Charles Bell, that the muscles
used in producing expression were created for this
special end.

In 1876 he re-commenced geological work, bringing
out his previous works on "Volcanic Islands," and on
"South America," as a single volume. In this year
too he wrote (November 16th) a most interesting
letter to James Geikie, offering an explanation of the
large stones standing in an upright position in the
drift of the south of England. He had noticed the
same thing with the flints in the red clay left upon the
chalk as a residuum after the action of solvent agencies
on the latter. This position he explained was due to
the movement following the slow subsidence of parts of
the clay as the chalk beneath dissolved, the flints arranging themselves along the lines of least resistance.
This suggested to him the view that the flints in the
drift are to be explained by the subsidence, during the
warmer climate which followed the glacial period, of
alternate layers of snow and drift accumulated during
the winters and summers respectively, of the cold
period itself.

This interesting view will, Geikie believes, come to
be accepted as the truth.

The book upon "The Formation of Vegetable Mould Through the Action of Worms," must be included among his geological works, although it contains a great many observations of deep zoological interest. It has been stated already that he wrote a paper on this subject for the Geological Society in.1838. In 1877 he studied the mode by which Roman remains gain their protective covering of mould ; again towards the end of 1880 he began systematically to prepare the book, which was published on October 10th of the following year. It was extremely successful, 8,500 copies being sold in three years.

This interesting work affords a good illustration of the tremendous results obtained, even in a moderate time, by an immense number of workers all using their powers in one direction. Each single earth-worm swallows earth in the excavation of its burrow and for the nutriment it contains, the waste material being ejected as "castings" at the surface, and as a lining to the burrow. But although the amount of earth thus swallowed by a single worm is not large, worms are so numerous that "the whole of the superficial mould . . . has passed, and will again pass, every few years through the bodies of worms." The result of this unceasing transport of the deeper mould to the surface is shown to be the burial of stones, either singly or in layers (as in paths), the covering and consequent protection of ancient buildings, and the preparation of soil for plants. In addition to this, the geological denuding agencies are assisted by the manner in which the deeper soil is brought into a position in which it is exposed to their action.

In 1879 he wrote and published a life of his grandfather, Erasmus Darwin, as " a preliminary notice " to the English translation of E. Krause's Life; but Darwin's contribution forms the larger part of the volume.

CHAPTER XXIV.

BOTANICAL WORKS (1862–86).

DARWIN's botanical works are referred to separately, and receive more systematic treatment than the others, in the great "Life and Letters." They form, together with the botanical letters, the subject of the seventh to the twelfth chapters in the last volume. It will therefore be unnecessary to treat them in any detail, although they form some of the most important and interesting of all his biological investigations.

Fertilisation of flowers.—"The Fertilisation of Orchids" was the first published of the botanical works, appearing in 1862, followed by a second and greatly altered edition in 1877. The object of the work "is to show that the contrivances by which orchids are fertilised are as varied and almost as perfect as any of the most beautiful adaptations in the animal kingdom; and secondly, to show that these contrivances have for their main object the fertilisation of the flowers with pollen brought by insects from a distant plant." Even in 1837 Darwin had written in his note-book, "Do not plants which have male and female organs together [*i.e.* in the same flower] yet receive influence from other plants? Does not Lyell give some argument about varieties being difficult to

M

keep [true] on account of pollen from other plants?
Because this may be applied to show all plants do
receive intermixture." (Quoted in the "Life and
Letters.") In 1841, Robert Brown, the distinguished
botanist, advised Darwin to read Sprengel's "Secret of
Nature Displayed" (Berlin, 1793). The result was to
encourage and assist Darwin in his work on fertilisation of flowers by insects, and to bring about the first
due recognition of Sprengel's merits, long after his
death.

"*The Effects of Cross- and Self-fertilisation in the
Vegetable Kingdom.*"—This work has a very direct
bearing on that last mentioned. Darwin speaks in
the Autobiography " of having come [in 1839] to the
conclusion in my speculations on the origin of species,
that crossing played an important part in keeping
specific forms constant." Later on he came to see
that the advantage of crossing is more direct, and
results from the greater vigour of the offspring over
those of self-fertilised plants. The object of this
work, published in 1876, was to prove this point by
experimental evidence of sufficient amount, and to
show in numerous cases, by measurements of height
or weight, or by counting the number of seeds produced, that cross-fertilisation invariably tends towards
the greater vigour of offspring.

Hence the motive cause for the marvellous
adaptations by which cross-fertilisation is ensured
was supplied.

"*Different Forms of Flowers on Plants of the Same*

Species" was published in 1877, and a second edition in 1880. This work, like so many others, had been largely anticipated by the author's original papers to scientific societies, in this case to the Linnean. The papers were combined, brought up to date, and with the addition of much new matter constituted the volume. The chief part of the work is concerned with heterostyled plants, viz. species which bear different kinds of flowers chiefly distinguished by the lengths of the pistil and stamens. As many as three different forms occur in Lythrum. In this work it is shown that each of the forms, although possessing both kinds of sexual organs, is adapted to be fertilised by the pollen of another form, and that such offspring are more vigorous than those produced by fertilisation by the same form. He furthermore showed that the offspring of " illegitimate " parentage (viz. those which were fertilised by the same form) possessed, in certain respects, a close resemblance to hybrids among animals. He remarks in his Autobiography, " No little discovery of mine ever gave me so much pleasure as the making out the meaning of heterostyled flowers."

In addition to the heterostyled flowers, the other differing forms borne by the same plants are considered, including the cleistogamic species, in which minute closed flowers are borne as well as the ordinary open ones. The former are wanting in the scents and colours of ordinary flowers, and are specially adapted for self-fertilisation, and the production of "an abundant supply of seeds with little expenditure."

"*Climbing Plants.*"—The subject of this volume was published as a paper before the Linnean Society in 1864. After being corrected, the material was brought out as a volume in 1875. Darwin, as he tells us in the Autobiography, was first led to study the subject by a paper by Asa Gray, which appeared in 1858 (Proc. Amer. Acad. of Arts and Sciences). Writing to Asa Gray, August 4th, 1863, he said, "My present hobby-horse I owe to you, viz. the tendrils." One of the most interesting results brought forward in this work is the fact that the upper growing part of a twining stem bends to one side and then travels slowly round, between two and three hours being required for each revolution, in the case of the hop growing in a room and observed at the period of most active movement. The circle swept at the 27th revolution was 19 inches in diameter. In the case of this plant the three youngest internodes (or joints), and never less than two of them, were concerned in the movement; "by the time the lower one ceased to revolve, the one above was in full action, with a terminal internode just commencing to move." The object of this movement is to strike some object round which the plant may twine. A much grander example was seen in *Ceropegia Gardnerii*, in which three long internodes and two short ones swept a circle over 5 feet in diameter, "at the rate of 32 or 33 inches per hour, in search of some object round which to twine." The stem of the plant is not in the least twisted by this movement. Nearly all of the great divisions of twining plants, leaf-climbers, and tendril-bearers "have the same remarkable power of spontaneously revolving."

"*The Power of Movements in Plants*" was published on November 6th, 1880. It embodies a vast amount of work carried on in conjunction with Francis Darwin. This volume bears a very direct relation to that last mentioned, as Darwin has explained in his Autobiography:—

"In accordance with the principle of evolution it was impossible to account for climbing plants having been developed in so many widely different groups unless all kinds of plants possess some slight power of movement of an analogous kind. This I proved to be the case ; and I was further led to a rather wide generalisation, viz. that the great and important classes of movements, excited by light, the attraction of gravity, &c., are all modified forms of the fundamental movement of circumnutation."

An extreme example of circumnutation has already been described in the revolving movements of the youngest parts of the stem of a twining plant.

The work evoked very great interest in this country, but was severely criticised by certain German botanists. The immense number of new observations must always have a very high value, whatever be the fate of the general conclusions, concerning which it may be remarked that Darwin's conclusions have often been criticised before, but time has shown that he was right.

" *Insectivorous Plants* " was published July 2nd, 1875, but I consider it last, as the subject stands somewhat apart from the rest of his botanical works. The subject was suggested to him by noticing the insects caught by the leaves of the Sun-dew (*Drosera*) near Hartfield. He then studied in great detail the causes of the movement, and the sensitiveness of the gland-tipped hairs, finding that a piece of hair weighing $\frac{1}{78000}$ of a grain causes one of them to curve inwards, and alters " the condition of the contents of every cell in the foot-stalk of the gland."

The greater part of the work deals with the experiments on *Drosera*, which were extremely numerous and detailed. The remainder treats of other insectivorous plants, such as Dionæa, Pinguicula, Utricularia, etc. The methods of capture, the movements of the plants under the stimulus supplied by the living insect (or other animal), and the resulting changes in the plant-cells were not the only points studied. He also investigated the digestive secretion and its action upon the food absorbed by the leaves.

CHAPTER XXV.

By the kindness of my friend Professor Meldola, and the courtesy of Mr. Francis Darwin, I am enabled to publish for the first time a series of letters written by Charles Darwin to the former. The whole series consists of 33 letters, written between January 28th, 1871, and February 2nd, 1882, only a few weeks before his death.

When we remember the immense amount of correspondence with which Darwin had to cope, the constant attention required by his investigations and publications, and the state of his health, it is deeply interesting to read these letters, written with such unfailing courtesy, to a younger worker in the lines that he had suggested, and who was thereby stimulated and encouraged to undertake the researches which are now so well known.

Reading these letters and remembering the circumstances of the writer, we can understand how it is that, although ill-health prevented his presence on occasions at which the younger scientific men are wont to meet—although he was known to but few of them—nevertheless the charm of his noble and generous nature was a most potent force in influencing and attracting men; and it was this, no less than his epoch-making discoveries, which has made it one of the chief regrets of many a scientific worker that he never saw Charles Darwin.

The correspondence was opened by a letter from Meldola informing Darwin of a case of hexadactylism in a man at Turnham Green.

"*Jan.* 28 [1871].   "Down.

"DEAR SIR,—I am much obliged for your kindness in informing me of the hexadactylous case; but so many have been recorded that I do not think, except under very special circumstances, it would be worth your while further to investigate it.

"With my thanks, yours faithfully and obliged,

"CH. DARWIN."

The next letter refers to Meldola's communication to *Nature* (he had recently written upon pangenesis and upon sexual selection), and his work on mimicry, protective resemblance, etc.   In the latter part we meet with an interesting reference to the researches on cross-fertilisation which are now so famous.

"*June 9th* [1871].   "Down.

"DEAR SIR,—I am greatly obliged by your note.   I have read with much interest and carefully perused your letter in Nature, and am looking out for a paper announced for Linn. Soc. Your remarks shall all be in due time fully considered.   With respect to the separation of the sexes, I have often reflected on the subject; but there is much difficulty, as it seems to me and as Nägeli has insisted, inasmuch as a strong case can be made out in favour of the view that with plants at least the sexes were primordially distinct, then became in many cases united, and in not a few cases re-separated. I have during the last 5 or 6 years been making a most laborious series of experiments, by which I shall be able, I think, to demonstrate the wonderful good derived from crossing, and I am almost sure but shall not know till the end of the summer that I shall be able to prove that the good is precisely of the same kind which the adult individual derives from *slight* changes of conditions.

"With my sincere thanks for your interest in my work, I remain, dear Sir, Yours very faithfully,

"CH. DARWIN."

The following letter is of great interest in relation to many problems of sexual selection, protective resemblance, mimicry, etc. :—

"*Jan.* 23, 1872. "Down.

"DEAR SIR—The point to which you refer seems to me a very difficult one. 1st the comparison of the amount of variability in itself would be difficult. 2ndly of all characters, colour seems to be the most variable, as we see in domesticated productions. (3) I fully agree that selection if long continued gives fixity to characters. We see the reverse of this in the great variability of fancy races, now being selected by man. But to give fixity, selection must be continued for a very long period : pray consider on this head what I have said in the Origin about the variability of characters developed in an extraordinary manner, in comparison with the same characters in allied species. The selection must also be for a definite object, and not for anything so vague as beauty, or for the superiority of one male in its weapons over another male, which can in like manner be modified. This at least seems to me partly to account for the general variability of secondary sexual characters. In the case of mimetic insects, there is another element of doubt, as the imitated form may be undergoing change which will be followed by the imitating form. This latter consideration seems to me, as remarked in my 'Descent of Man,' to throw much light on how the process of imitation first began.

"I enclose a letter from Fritz Müller which I think is well worth reading, and which please to return to me.

"You will see he lays much stress on the difficulty of several remotely allied forms all imitating some one species. Mr. Wallace did not think that there was so much weight in this objection as I do. It is, however, possible that a few species in widely different groups, before they had diverged much, should have accidentally resembled, to a certain extent, some one species. You will also see in this letter a strange speculation, which I should not dare to publish, about the appreciation of certain colours being developed in those species which frequently behold other forms similarly ornamented. I do not feel at all sure that this view is as incredible as it may at first appear. Similar ideas have passed through my mind when considering the dull colours of all the organisms which inhabit dull-coloured regions, such as Patagonia and the Galapagos Is. I suppose you know Mr. Riley's excellent essay on mimicry in the last report on the noxious insects of Missouri or some such title.

"I hope your work may be in every way successful.

"I remain, dear Sir, yours faithfully,

"CHARLES DARWIN."

The next letter deals with mimetic resemblance :—

"*Mar.* 28, 1872.   "Down.

"DEAR SIR—I thank you for your information on various subjects. The point to which you allude seems to me very obscure, and I hardly venture to express an opinion on it My first impression is that the colour of an imitating form might be modified to any extent without any tendency being given to the retention of ancient structural peculiarities. The difficulty of the subject seems to me to follow from our complete ignorance of the causes which have led to the generic differences between the imitating and imitated forms. The subject however seems worth investigating. If the imitator habitually lives in company with the imitated, it would be apt to follow in some respects the same habits of life, and this perhaps would lead to the retention or acquirement of some of the same structural characters.

"I wish you all success in your essay, and remain, dear Sir, yours very faithfully, "CH. DARWIN."

The next very brief letter, acknowledging the receipt of a note, was written from Down, March 26th, 1873. It contained some sympathetic remarks upon the progress of Meldola's work upon Mimicry. In the succeeding letter, printed below, we find a very definite statement of opinion as to the *rôle* of monstrosities in evolution :—

"*Aug.* 13*th* [1873]. "Down.

"DEAR SIR—I am much obliged for your present which no doubt I shall find at Down on my return home. . . .

"I am sorry to say that I cannot answer your question ; nor do I believe that you could find it anywhere even approximately answered. It is very difficult or impossible to define what is meant by a larger variation. Such graduate into monstrosities or generally injurious variations. I do not myself believe that these are often or ever taken advantage of under nature. It is a common occurrence that *abrupt* and considerable variations are transmitted in an unaltered state, or not at all transmitted, to the offspring or to some of them. So it is with tailless or hornless animals, and with sudden and great changes of colour in flowers.—I wish I could have given you any answer.

"Dear Sir, yours very faithfully, "CH. DARWIN."

The succeeding three letters show Darwin's scrupulous care as regards the publication, although with every acknowledgment, of the results obtained by others. They refer to a letter from Fritz Müller which he had forwarded to Meldola. The latter had written to ask Darwin's permission and advice as to the inclusion of some of F. Müller's observations in his most interesting paper, "Entomological Notes bearing on Evolution" (*Ann. and Mag. Nat. Hist.*, 1878, 5th series, Vol. I. p. 155), which he was then preparing :—

"*Sept.* 14, 1877.   "Down.

"DEAR SIR—I have some doubts whether Fritz Müller would like extracts from his letters being published after so long an interval,—that is if the letter relates to the origin of mimicry ; for he published about a year ago an excellent paper on this subject. I believe it was in the Jenaische Zeitschrift, but the paper is out of its proper place in my library and I cannot find it. If you thought it worth while to send me your copy I could then judge about the publication of extracts.

"I fear it is not likely that I shall have anything to communicate to the Entomological Soc. I quite agree with you that it is a great pity that our Entomologists should confine themselves to describing species.

"Dear Sir, yours faithfully,   "CH. DARWIN."

"*Sept.* 22*nd* [1877].   "Down.

"MY DEAR SIR—I am doubtful whether speculations in a letter ought to be published, especially after a long interval of time. Any fact which he states, I feel pretty sure he would not at all object being used by anyone.—Pray do the best you can.—I should grieve beyond measure to be accused of a breach of confidence.—He has lately, as I mentioned, thrown much light on the first steps in mimickry.

"With respect to dimorphic Butterflies, those about which I have read appear at different seasons, and have been the subject of an *admirable* essay by Prof. Weismann. It is some little time since I read the essay and one subject drives another out of my head, but I think he explains all such cases by the direct inherited effects of temperature. He tried experiments. If you read German, I believe I could find Weismann's essays and lend them to you. In your present interesting case I really do not know what to think : it seems rather bold to attribute the 2 coloured forms to nat. selection, before some advantage can be pointed out.—May not the female revert in some cases ? I do not doubt that the intermediate form could be eliminated as you suggest.

"I wish that my opinion could have been of any value. . . .

"I remain yours very faithfully,   "CH. DARWIN."

This last letter, with others that followed it, directing Meldola's attention to Weismann's "Studies in the Theory of Descent," resulted in the English translation which is so admirably rendered and edited. Many of the later letters are concerned with the progress of this publication. The remarks about dimorphic butterflies referred to Meldola's observation, that in one of those years in which *Colias edusa* was extremely abundant, a whole series of forms had been taken transitional between the normal orange female and the white variety *helice* :—

"*Sept.* 27 [1877].   "Down.

"MY DEAR SIR,—It is impossible for F. M. [Fritz Müller] to object to anything which you have said in your very interesting little essay.—I just allude to Butterflies preferring certain colours at p. 317 of 2nd Editⁿ of the Descent and to the case of the species of Castnia p. 315 which has ornamented hinder wings and displays them, whilst 2 other species have plain hind wings and do not display them. My son, who has charge of my library, returns home to-night and then we will search for Weismann. He gives splendid case of caterpillar with coloured ocelli like true eyes, *and which frightened away birds.*

"Yours sincerely,   "CH. DARWIN."

The reference in this letter is to Meldola's paper, "Entomological Notes bearing on Evolution," soon afterwards published in the *Annals and Magazine of Natural History*, 1878, Vol. I. p. 155. The caterpillar referred to is the well-known larva of the Large Elephant Hawk Moth (*Chœrocampa elpenor*).

Darwin then wrote a brief note (October 19th, 1877) referring to a number of *Kosmos* containing an article on " Sexual Selection." He offered to send the number if it would interest his correspondent. The number was sent, as the succeeding letter shows :—

" *Oct. 22nd* [1877].   " Down.

' MY DEAR SIR,—I send Kosmos by this post. . . .

" Prof. Weismann's address is Freiburg.—I should think he would be glad of translation, and would probably arrange for stereotypes of Plates.—You could say as an introduction that I had lent you his book.—To find a publisher will be perhaps a difficulty. Should it be translated I must beg you to get another copy, as I cannot spare mine for such a length of time.—Wallace sent me his article and I was quite dissatisfied with it.—To explain a peacock's tail by vital activity seems to me mere verbiage—a mere metaphysical principle.

" My dear Sir, yours faithfully,   " CH. DARWIN.

" It will be a public benefit to bring out a translation."

Then followed three letters, January 3rd, March 24th, and March 27th, 1878 ; the first written when Darwin was sending another number of *Kosmos*, the second when sending his photograph, the third enclosing a letter from Fritz Müller containing some very interesting observations on mimicry in South American butterflies.

He then wrote as follows :—

"*April* 17/78.   "Down.

"My dear Sir,—I should be very much obliged if you could get some one to name the photographs of the enclosed insect and read the enclosed letter. It seems a pretty, but I think not new case of protective resemblance. One might fancy that the large ocelli on the under wings were a sexual ornament.—Perhaps these photographs might be worth exhibiting at the Entomolog. Soc.—I do not want them returned (unless indeed Dr. Zacharias wants them back, which is not probable) or the enclosed letter.

"A single word with the name of the genus and if possible of the species, would suffice.—

"Pray forgive my troubling you and believe me

"Yours faithfully,   "Ch. Darwin.

"I am glad that F. Müller's letter interested you. He has published a paper with plates on the shape of the hairs or scales on the odoriferous glands of many butterflies, which I could send you, but I doubt whether you would care for it."

Darwin then sent another letter from Fritz Müller containing some interesting notes on odoriferous organs in butterflies, and on the occasional failure of the female insect to deposit her eggs on a plant which can serve as the food of the young larvæ. The beetles alluded to were a species of *Spermophagus*. The two letters printed below refer to the same subjects :—

"*May* 15 [1878].   "Down.

"My dear Sir,—I think the enclosed will interest you.— The letter to me need not be returned as I have had the only important passage for my work copied out.—In the letter F. M. [Fritz Müller] sent me seeds of *Cassia neglecta* and several beetles arrived alive, having formed their cocoons, and gnawed their way out of the little peas or seeds.—These elegant beetles, with the knowledge of their manner of development may interest some Coleopterist.

" I hope to hear some time about Dr. Zacharias' photographs. I received your obliging letter from Paris.

"Yours sincerely,   "Ch. Darwin."

" *May* 25 [1878].   "Down.

" My dear Sir,—The living beetles and the cocoons were found in a small paper packet containing the seeds. Those from which the beetles had emerged were much broken, and the larvæ had evidently attacked some of the other seeds. I am sorry to say that some of the injured ones were thrown away. I am glad that you are going to draw up a paper from Fritz Müller's letters.

"Yours sincerely,   "Ch. Darwin."

After another short note, dated July 24th, 1878, Darwin wrote the following letter, which explains how it was that he came to write the preface to the translation of Weismann's "Studies":—

" *October* 31 [1878].   "Down.

" My dear Sir,—As you are inclined to be so very liberal as to have a translation made of Weismann's Essays on your own risk, I feel bound to aid you to the small extent of writing a short prefatory notice. But this is a kind of job, which I do not feel that I can do at all well and therefore do not like ; but I will do my best. It must, however, be short for I am at present working very hard. I do not quite understand whether you intend asking some Publisher to bring out the book on commission at your cost for if so there will be no difficulty in finding a Publisher. But if you expect any Publisher to publish at his risk and cost; I think from recent experience you will have much difficulty in finding one.—I suppose that you have asked Weismann's concurrence.

" Down is rather an awkward place to reach, as we are

4 miles from nearest station, Orpington. But I shall be in
London for a week on Novr 17th or 18th and could see you
then at any time, and perhaps you could come to luncheon.

"But if you would prefer to come here, I shall be very
happy to see you either Saturday or Sunday, if you would let
me know hour.—I am, however, bound to tell you that my
health is always doubtful, and that my head does not allow
me to converse long with anyone.

"With the most cordial sympathy in your undertaking, I
remain, my dear Sir, yours very faithfully, "CH. DARWIN."

In November, 1878, Darwin was in London,
staying at his daughter's house at 4, Bryanston
Street. On the 19th he wrote asking Meldola to
lunch to talk over the proposed English edition of
Weismann, and on the 25th sent the MS. of the
Preface with the following letter:—

"4 Bryanston St.,
"*Nov.* 25 [1878]. "Portman Sq^{re}.

"MY DEAR SIR,—I send my little Preface, which I do not
at all like, but which I cannot improve. I should like hereafter to see it in type. Mr. Bates tells me that Hardwick and
Bohn of Piccadilly intend to go in for publishing solid books;
and if your present publisher should change his mind: Mr.
Bohn might be worth applying to.

"Yours sincerely, "CH. DARWIN."

Professor Meldola then wrote, suggesting that
Darwin should, in his Preface, point out, by references
to the "Origin of Species" and his other writings,
how far he had already traced out the lines which
Weismann had pursued in his researches. The
suggestion was made because in a great many of
the Continental writings upon the theory of descent

N

a number of the points which had been clearly fore-shadowed, and in some cases even explicitly stated, by Darwin had been independently rediscovered and published as though original. In the editorial notes to Meldola's translation full justice to Darwin has been done in this respect. Darwin's characteristic reply is deeply interesting.

"*Nov* 26th [1878].   " 4 Bryanston St.

"My dear Sir,—I am very sorry to say that I cannot agree to your suggestion.—An author is never a fit judge of his own work, and I should dislike extremely pointing out when and how Weismann's conclusions and work, agreed with my own. — I feel sure that I ought not to do this, and it would be to me an intolerable task. Nor does it seem to me the proper office of the Preface, which is to show what the book contains and that the contents appear to me valuable. But I can see no objection for you, if you think fit, to write an introduction with remarks or criticisms of any kind. Of course I would be glad to advise you on any point as far as lay in my power, but as a whole I could have nothing to do with it, on the grounds above specified that an author cannot and ought not to attempt to judge his own works or compare them with others. I am sorry to refuse to do anything which you wish.—

"We return home early to-morrow morning.—Your green silk seems to me a splendid colour, whatever the æsthetics may say.—My dear Sir, yours faithfully,   "Ch. Darwin."

The "green silk" referred to some specimens of coal-tar colours sent to show Darwin what modern chemistry had been able to accomplish in the way of artificial colouring matters. They were at that time of particular interest in connection with a discussion which had arisen in Bryanston Street about the so-called "æsthetic" school, which had become rather predominant at the period, and which affected an abhorrence of all brilliant colouring, in spite of the circumstance that nature abounds in the most gorgeous hues, especially in the tropics.

The next letter refers to the adoption of the word "phyletic" in the translation of Weismann.

"*Dec.* 14 [1878]. "Down.

"MY DEAR SIR,—I am very glad that you are making good progress with the book.—You could not apply to a worse person than myself on any philological question. I presume that 'phyletische' has been adopted or modified from Häckel. As the latter uses the word, it has nearly the sense of genealogical. It always applies to the lines of descent, and therefore differs somewhat from 'innate'; for an inherited character, though derived from the father alone or only a single generation, would be innate in the child. I should think 'phyletic' would do very well, if you gave the German word and an explanation, in a foot-note.

"There has been a delay in answering your letter, but I have just heard from my son who is away from home, and he says that he is sorry but he cannot well spare the time to lecture.

"My dear Sir, yours very faithfully,
"CH. DARWIN."

Then followed two letters (January 20th, and February 7th, 1879), the first written when Darwin was sending a number of *Kosmos;* the second referring to it and other papers, and asking that his name should be put down as a subscriber to the forthcoming translation of Weismann.

Later on the number of *Kosmos* for May, 1879, was sent, containing (p. 100) Fritz Müller's paper "*Ituna* and *Thyridia.*" This paper, although it did not attract sufficient attention at the time, was of the highest importance in relation to the theory of mimicry, as Meldola at once perceived.

Bates in his epoch-making paper in the Transactions of the Linnean Society (Vol. XXIII. 1862) had founded the theory of mimicry. Those rarer forms which have diverged from their near allies and, in superficial appearance, approached some distantly related, but abundant, species inhabiting the same tract have been, according to Bates's theory, benefiting themselves in the struggle for existence. The mimicked species are, he suggested, abundant because they possess some special means of protection, such as an unpleasant taste or smell, and they have an unpleasant reputation which greatly aids them in the struggle for life; while the mimicking species, by their superficial resemblance, are enabled to live upon that reputation without possessing the special means of defence.

Certain facts well-known to Bates, and brought forward in his paper, were not explicable by this theory, viz. the resemblance that often exists between the abundant and specially protected species themselves. Although a few tentative suggestions were made, such as the production of a common appearance by similarity of climate, or food, etc., these facts remained an unexplained mystery until this paper of Fritz Müller's in the May number of *Kosmos.* He there suggests that the mutual resemblance between the specially protected forms is advantageous, in reducing for each of them the number of individuals which must be sacrificed during the process of education which their youthful enemies must undergo, before they learn what is fit and what unfit for food. The arrangement is, in fact, much like that between a couple of firms that issue a common advertisement, and so save about half the expense of advertising alone. It is only another added to the numerous examples of the production by natural selection, and without the introduction of consciousness, of a result which could not be bettered by the deliberate action of the most acute intelligence.

.Meldola at once wrote to Darwin asking his advice about the translation of F. Müller's paper, and received the following reply :—

"*June 6th*, 1879.   "Down.

"My dear Mr. Meldola,—Your best plan will be to write to ' Dr. Ernst Krause, Friedenstrasse, 10 II. Berlin.' He is one of the editors with whom I have corresponded. You can say that I sent you the Journal and called your attention to the paper, but I cannot take the liberty of advising the supply of clichés. He is a very obliging man. Had you not better ask for permission to translate, saying the source will be fully acknowledged ?

" F. Müller's view of the mutual protection was quite new to me..

"Yours sincerely,   "Ch. Darwin."

The clichés were obtained and Meldola's translation published in the Proceedings of the Entomological Society for 1879, p. 20. The new contribution to the theory of mimicry was at first somewhat severely criticised, even Bates being adverse to it. Subsequent work has abundantly justified it as by far the most important addition to the subject since Bates's original paper. In fact, many cases which have been up to the present explained under the theory of true (Batesian) mimicry are now believed to come under that which we owe to F. Müller—viz. convergence between specially protected forms for mutual benefit.

An interesting paper by Dr. F. A. Dixey, published in the Transactions of the Entomological Society for the present year (1896), contains convincing arguments in favour of this view as regards some of the *Pieridæ* of South America in relation to the *Heliconidæ* and *Papilionidæ* which they resemble.

It is of the highest interest to learn that the first introduction of this new and most suggestive hypothesis into this country was due to the direct influence of Darwin himself, who brought it before the notice of the one man who was likely to appreciate it at its true value and to find the means for its presentation to English naturalists.

In the next year Meldola wished to translate further papers of Fritz Müller's, and received the following letter on the subject:—

"*Nov.* 25/80.   "Down.

"My DEAR SIR,—I can well believe that your labour must have been great, and everyone is bound to aid you in any way.

"No. I. of F. Müller's paper is in the August no. for 1877.

" No. II.—is in the October no. for 1877.

" Both these articles I remember thinking excellent.

" I am not one of the editors of Kosmos, only a kind of patron (!) and therefore cannot give permission ; but when you write to the editors you can say that I have expressed a hope that permission would be granted, you acknowledging source of papers.

" Heartily wishing you success and in haste to catch first post, I remain yours very faithfully,   " CH. DARWIN."

Shortly after the date of the last letter Professor Meldola came across a copy of Thomson's " Annals of Philosophy" on a bookstall. It bore the name " Erasmus Darwin" on the first page, and Meldola offered it to Charles Darwin, thinking it might have belonged to his grandfather.

" *March* 12*th*, 1881 [The date was evidently May, and not March].   " Down.

" DEAR MR. MELDOLA,—It is very kind of you to offer to send me the book, but I feel sure that it could not have belonged to my grandfather.—My eldest brother's name is Erasmus and he attended to chemistry when young, and I suppose that the 'Annals of Philosophy' was left at my Father's house and sold with the Library which belonged to my sisters.—I will look to the few words of Preface to Wiesmann [*sic*], whenever I receive a proof.—With many thanks.—

  " Yours very faithfully,   " CH. DARWIN."

Then followed a brief note dated " Aug. 8, 1881," referring to some point in the work upon which Meldola was then engaged, and which cannot now be ascertained. Another letter of the same date referred to the translation of Weismann, and contained some encouraging words upon the interest created by the work and upon the success of the Essex Field Club, in which Meldola had taken a leading part. Another brief note of August 10th, 1881, apparently refers to some paper which cannot now be identified.

The following interesting letter is of uncertain date :—

    "? 19*th*, ? 1881.   "Down.

"DEAR MR. MELDOLA,—When I read the F. M. [Fritz Müller] paper your doubt occurred to me and I must say this, I would rather have expected that the knowledge of distasteful caterpillars would have been inherited, but I distinctly remember an account (when Wallace first propounded his warning colors) published of some birds, I think turkeys, being experimented upon and they shook their heads after trying some caterpillars as if they had a horrid taste in their mouths. I fancied this thing was published by Mr. Weir or could it have been by Mr. Butler? It would be well to look in Mr. Belt's 'Nicaragua' as he tried some experiments. I am not sure that there is not some statement of the kind in it.

          "Yours faithfully,   "CHARLES DARWIN.

"I daresay Mr. Wallace or Bates would remember the statement of some birds shaking their heads to which I refer."

The statement about the turkeys evidently refers to Stainton's experiment with young birds of this kind, which immediately devoured numerous protectively coloured moths, but, after seizing, invariably rejected, a conspicuous white species (*Spilosoma menthastri*). It was Belt's ducks which shook their heads after tasting a very conspicuous Nicaraguan frog. Darwin wished to show by this evidence that there was no instinctive knowledge such as would have saved the birds from an evidently unpleasant experience.

The last letter, deeply interesting both on its own account and because it was written so near the end of Darwin's life, was a reply to one from Meldola in which he had said that the publishers were complaining that the list of subscribers was disappointing, and that they had expressed the wish that Mr. Darwin could see his way to writing a much longer introductory notice than he had done.

"*Feb. 2nd* [1882].   "Down.

"DEAR MR. MELDOLA,—I am very sorry that I can add nothing to my very brief notice without reading again Weismann's work and getting up the whole subject by reading my own and other books, and for so much labour I have not strength. I have now been working at other subjects for some years, and when a man grows as old as I am, it is a great wrench to his brain to go back to old and half-forgotten subjects. You would not readily believe how often I am asked questions of all kinds, and quite lately I have had to give up much time to do a work, not at all concerning myself, but which I did not like to refuse. I must however somewhere draw the line, or my life will be a misery to me.

"I have read your Preface and it seems to me *excellent.* I am sorry in many ways, including the honour of England as a scientific country, that your translation has as yet sold badly. Does the publisher or do you lose by it ? If the publisher, though I should be sorry for him, yet it is in the way of business ; but if you yourself lose by it, I earnestly beg you to allow me to subscribe a trifle, viz. ten guineas, towards the expense of this work, which you have undertaken on public grounds.

"Pray believe me, yours very faithfully,

"CH. DARWIN."

Darwin's generous offer, although gratefully declined, was a warm encouragement in the laborious,
and in some respects thankless, task of translator and
editor—a task which, in the case of the English edition
of Weismann's "Studies in the Theory of Descent,"
was carried out in so admirable a manner.

CHAPTER XXVI.

HIS LAST ILLNESS (1882).

In the last few months of his life, towards the end of 1881 and beginning of 1882, Darwin began to suffer from his heart, causing attacks of pain and faintness which increased in number. On March 7th, 1882, he had one of these seizures when walking, " and this was the last time that he was able to reach his favourite 'sand-walk'" ("Life and Letters "). After this he became rather better, and on April 17th was able to record the progress of an experiment for his son Francis. The following sentences are quoted from the " Life and Letters ":—

" During the night of April 18th, about a quarter to twelve, he had a severe attack and passed into a faint, from which he was brought back to consciousness with very great difficulty. He seemed to recognise the approach of death, and said, 'I am not the least afraid to die.' All the next morning he suffered from terrible nausea and faintness, and hardly rallied before the end came.

" He died at about four o'clock on Wednesday, April 19th, 1882."

He was buried in Westminster Abbey on April 26th.

Thus died one of the greatest of men, after a life of patient and continuous work interrupted only by ill-health; a man who was, perhaps, more widely attacked and more grossly misrepresented than any other, but who lived to see his teachings almost universally received; a man whose quiet, peaceful life of work, and whose precarious health, prevented that large intercourse with his fellow-men which is generally forced upon greatness, but who was so beloved by his circle of intimate friends that, through their contagious enthusiasm, and through the glimpses of his nature revealed in his writings, he was in all likelihood more greatly loved than any other man of his time by those who knew him not.

And for all those of ushw o have loved Darwin, although we have never seen him, we can at any rate remember that we have lived in his time and have heard the echoes of his living voice; he has been even more to us than he will be to future generations of mankind—a mighty tradition, gaining rather than losing in force and in overwhelming interest as each passing age, inspired by his example, guided by his teachings, adds to the knowledge of nature, and in so doing gives an ever deeper meaning to his life and work.

INDEX.

PRINTED BY CASSELL & COMPANY, LIMITED, LA BELLE SAUVAGE, LONDON, E.C.

Illustrated, Fine-Art, and other Volumes.

Adventure, The World of. Fully Illustrated. In Three Vols. 9s. each.

Adventures in Criticism By A. T. QUILLER-COUCH. 6s.

Africa and its Explorers, The Story of. By DR. ROBERT BROWN, F.L.S. Illustrated. Complete in 4 Vols., 7s. 6d. each.

Animals, Popular History of. By HENRY SCHERREN, F.Z.S. With 13 Coloured Plates and other Illustrations. 7s. 6d.

Architectural Drawing. By R. PHENÉ SPIERS. Illustrated. 10s. 6d.

Art, The Magazine of. Yearly Vol. With 14 Photogravures or Etchings, a Series of Full-page Plates, and about 400 Illustrations. 21s.

Artistic Anatomy. By Prof. M. DUVAL. *Cheap Edition.* 3s. 6d.

Astronomy, The Dawn of. By Prof. J. NORMAN LOCKYER, C.B., F.R.S., &c. Illustrated. 21s.

Atlas, The Universal. A New and Complete General Atlas of the World. List of Maps, Prices and all Particulars on Application.

Ballads and Songs. By WILLIAM MAKEPEACE THACKERAY. With Original Illustrations. 6s.

Barber, Charles Burton, The Works of. With Forty-one Plates and Portraits, and Introduction by HARRY FURNISS. 21s. net.

Battles of the Nineteenth Century. An entirely New and Original Work, with several hundred Illustrations. In 2 Vols., 9s. each

Beetles, Butterflies, Moths, and Other Insects. By A. W. KAPPEL, F.E.S., and W. EGMONT KIRBY. With 12 Coloured Plates. 3s. 6d.

"Belle Sauvage" Library, The. Cloth, 2s. each. A list of the Volumes post free on application.

Biographical Dictionary, Cassell's New. *Cheap Edition,* 3s. 6d.

Birds' Nests, British: How, Where, and When to Find and Identify Them. By R. KEARTON. With an Introduction by Dr. BOWDLER SHARPE and nearly 130 Illustrations of Nests, Eggs, Young, etc., from Photographs by C. KEARTON. 21s.

Birds' Nests, Eggs, and Egg-Collecting. By R. KEARTON. Illustrated with 22 Coloured Plates. *Fourth and Enlarged Edition.* 5s.

Black Watch, The. A Vivid Descriptive Account of this Famous Regiment. By ARCHIBALD FORBES. 6s.

Britain's Roll of Glory; or, the Victoria Cross, its Heroes, and their Valour. By D. H. PARRY. Illustrated. 7s. 6d.

British Ballads. With Several Hundred Original Illustrations. Complete in Two Vols., cloth, 15s. Half morocco, *price on application.*

British Battles on Land and Sea. By JAMES GRANT. With about 800 Illustrations. Four Vols., 4to, £1 16s.; *Library Edition,* £2.

Building World. Half-yearly Vols., I. and II., 4s. each.

Butterflies and Moths, European. With 61 Coloured Plates. 35s.

Canaries and Cage-Birds, The Illustrated Book of. With 56 Facsimile Coloured Plates, 35s. Half-morocco, £2 5s.

Captain Horn, The Adventures of. By FRANK STOCKTON. 6s.

Capture of the "Estrella," The. A Tale of the Slave Trade. By COMMANDER CLAUDE HARDING, R.N. *Cheap Illustrated Edition.* 3s. 6d.

Cassell's Family Magazine. Yearly Vol. Illustrated. 7s. 6d.

Cathedrals, Abbeys, and Churches of England and Wales. Descriptive, Historical, Pictorial. *Popular Edition.* Two Vols. 25s.

Cats and Kittens. By HENRIETTE RONNER. With Portrait and 13 Full-page Photogravure Plates and numerous Illustrations. £2 2s.

Chums. The Illustrated Paper for Boys. Yearly Volume, 8s.

Cities of the World. Four Vols. Illustrated. 7s. 6d. each.

Civil Service, Guide to Employment in the. Entirely New Edition. Paper, 1s. Cloth, 1s. 6d.

Clinical Manuals for Practitioners and Students of Medicine. A List of Volumes forwarded post free on application to the Publishers.

Colour. By Prof. A. H. CHURCH. With Coloured Plates. 3s. 6d.

Conning Tower, In a; or How I Took H.M.S. " Majestic" into Action. By H. O. ARNOLD-FORSTER, M.P. *Cheap Edition.* Illd. 6d.

Cook, The Thorough Good. By GEORGE AUGUSTUS SALA. With 900 Recipes. 21s.

Cookery, Cassell's Dictionary of. With about 9,000 Recipes, and Key to the Principles of Cookery. 5s.

Cookery, A Year's. By PHYLLIS BROWNE. 3s. 6d.

Cookery Book, Cassell's New Universal. By LIZZIE HERITAGE. With 12 Coloured Plates and other Illustrations. Strongly bound in Half-leather. 1,344 pages. 6s.

Cookery, Cassell's Shilling. 125*th Thousand.* 1s.

Cookery, Vegetarian. By A. G. PAYNE. 1s. 6d.

Cooking by Gas, The Art of. By MARIE J. SUGG. Illustrated. 2s.

Cottage Gardening, Poultry, Bees, Allotments, Etc. Edited by W. ROBINSON. Illustrated. Half-yearly Volumes, 2s. 6d. each.

Countries of the World, The. By ROBERT BROWN, M.A., Ph.D., &c. *Cheap Edition.* Illustrated. Vols. I., II., and III., 6s. each.

Cyclopædia, Cassell's Concise. Brought down to the latest date. With about 600 Illustrations. *Cheap Edition.* 7s. 6d.

Cyclopædia, Cassell's Miniature. Containing 30,000 subjects. Cloth, 2s. 6d. ; half-roxburgh, 4s.

David Balfour, The Adventures of. By R. L. STEVENSON. Illustrated. Two Vols 6s. each.
Part 1.—Kidnapped.   Part 2.—Catriona.

Diet and Cookery for Common Ailments. By a Fellow of the Royal College of Physicians, and PHYLLIS BROWNE. *Cheap Edition,* 2s. 6d.

Dog, Illustrated Book of the. By VERO SHAW, B.A. With 28 Coloured Plates. Cloth bevelled, 35s. ; half-morocco, 45s.

Domestic Dictionary, The. Illustrated. Cloth, 7s. 6d.

Doré Bible, The. With 200 Full-page Illustrations by DORÉ. 15s.

Doré Don Quixote, The. With about 400 Illustrations by GUSTAVE DORÉ. *Cheap Edition.* Bevelled boards, gilt edges, 10s. 6d.

Doré Gallery, The. With 250 Illustrations by DORÉ. 4to, 42s.

Doré's Dante's Inferno. Illustrated by GUSTAVE DORÉ. With Preface by A. J. BUTLER. Large 4to Edition, cloth gilt, 21s. Cloth gilt or buckram, 7s. 6d.

Doré's Dante's Purgatory and Paradise. Illustrated by GUSTAVE DORÉ. *Cheap Edition.* 7s. 6d

Doré's Milton's Paradise Lost. Illustrated by DORÉ. 4to, 21s. *Popular Edition.* Cloth gilt or buckram gilt, 7s. 6d.

Earth, Our, and its Story. By Dr. ROBERT BROWN, F.L.S. With Coloured Plates and numerous Wood Engravings. Three Vols. 9s. each.

Edinburgh, Old and New. With 600 Illustrations. Three Vols. 9s. each.

Egypt: Descriptive, Historical, and Picturesque. By Prof. G. EBERS. With 800 Original Engravings. *Popular Edition.* In Two Vols. 42s.

Electric Current, The. How Produced and How Used. By R. MULLINEUX WALMSLEY, D.Sc., etc. Illustrated. 10s. 6d.

Electricity in the Service of Man. Illustrated. *New and Revised Edition.* 10s. 6d.

Electricity, Practical. By Prof. W. E. AYRTON. 7s. 6d.

Encyclopædic Dictionary, The. In Fourteen Divisional Vols., 10s. 6d. each; or Seven Vols., half-morocco, 21s. each ; half-russia, 25s.

England, Cassell's Illustrated History of. With upwards of 2,000 Illustrations. *Revised Edition.* Complete in Eight Vols., 9s. each ; cloth gilt, and embossed gilt top and headbanded, £4 net the set.

English Dictionary, Cassell's. Giving definitions of more than 100,000 Words and Phrases. *Superior Edition,* 5s. *Cheap Edition,* 3s. 6d.

English History, The Dictionary of. Edited by SIDNEY LOW, B.A., and Prof. F. S. PULLING, M.A. *Cheap Edition*, 10s. 6d.
English Literature, Library of. By Prof. HENRY MORLEY. Complete in Five Vols., 7s. 6d. each.
English Literature, The Dictionary of. By W. DAVENPORT ADAMS. *Cheap Edition.* 7s. 6d.
English Literature, Morley's First Sketch of. *Revised Edition.* 7s. 6d.
English Literature, The Story of. By ANNA BUCKLAND. 3s. 6d.
English Writers. By Prof. HENRY MORLEY. Vols. I. to XI. 5s. each.
Etiquette of Good Society. *New Edition.* Edited and Revised by LADY COLIN CAMPBELL. 1s. ; cloth, 1s. 6d.
Fairway Island. By HORACE HUTCHINSON. *Cheap Edition.* 3s. 6d.
Fairy Tales Far and Near. Re-told by Q. Illustrated. 3s. 6d.
Fiction, Cassell's Popular Library of. 3s. 6d. each.

THE SQUIRE. By MRS. PARR.
THE AWKWARD SQUADS, and Other Ulster Stories. By SHAN F. BULLOCK.
THE AVENGER OF BLOOD. By J. MACLAREN COBBAN.
A MODERN DICK WHITTINGTON. By JAMES PAYN.
THE MAN IN BLACK. By STANLEY WEYMAN.
A BLOT OF INK. Translated by Q. and PAUL M. FRANCKE.
THE MEDICINE LADY. By L. T. MEADE.
OUT OF THE JAWS OF DEATH. By FRANK BARRETT.

IA. A LOVE STORY. By Q.
PLAYTHINGS AND PARODIES. Short Stories and Sketches. By BARRY PAIN.
A KING'S HUSSAR. By H. COMPTON.
"LA BELLA" AND OTHERS. By EGERTON CASTLE.
LEONA. By Mrs. MOLESWORTH.
FOURTEEN TO ONE, ETC. By ELIZABETH STUART PHELPS.
FATHER STAFFORD. By ANTHONY HOPE.
DR. DUMÁNY'S WIFE. By MAURUS JÓKAI.
THE DOINGS OF RAFFLES HAW. By CONAN DOYLE.

Field Naturalist's Handbook, The. By the Revs. J. G. WOOD and THEODORE WOOD. *Cheap Edition.* 2s. 6d.
Figuier's Popular Scientific Works. With Several Hundred Illustrations in each. Newly Revised and Corrected. 3s. 6d. each.
THE HUMAN RACE. MAMMALIA. OCEAN WORLD.
THE INSECT WORLD. REPTILES AND BIRDS.
WORLD BEFORE THE DELUGE. THE VEGETABLE WORLD.
Flora's Feast. A Masque of Flowers. Penned and Pictured by WALTER CRANE. With 40 Pages in Colours. 5s.
Football, The Rugby Union Game. Edited by REV. F. MARSHALL. Illustrated. *New and Enlarged Edition.* 7s. 6d.
For Glory and Renown. By D. H. PARRY. Illustrated. 3s. 6d.
France, From the Memoirs of a Minister of. By STANLEY WEYMAN. 6s.
Franco-German War, Cassell's History of the. Complete in Two Vols. Containing about 500 Illustrations. 9s. each.
Free Lance in a Far Land, A. By HERBERT COMPTON. 6s.
Garden Flowers, Familiar. By SHIRLEY HIBBERD. With Coloured Plates by F. E. HULME, F.L.S. Complete in Five Series. 12s. 6d. each.
Gardening, Cassell's Popular. Illustrated. Four Vols. 5s. each.
Gazetteer of Great Britain and Ireland, Cassell's. Illustrated. Vols. I. II. and III. 7s. 6d. each.
Gleanings from Popular Authors. Two Vols. With Original Illustrations. 4to, 9s. each. Two Vols. in One, 15s.
Gulliver's Travels. With 88 Engravings by MORTEN. *Cheap Edition.* Cloth, 3s. 6d. ; cloth gilt, 5s.
Gun and its Development, The. By W. W. GREENER. With 500 Illustrations. 10s. 6d.
Heavens, The Story of the. By Sir ROBERT STAWELL BALL, LL.D., F.R.S., F.R.A.S. With Coloured Plates. *Popular Edition.* 12s. 6d.
Highway of Sorrow, The. By HESBA STRETTON and ******** 6s.

Hispaniola Plate (1683-1893). By JOHN BLOUNDELLE-BURTON. 6s.

History, A Foot-note to. Eight Years of Trouble in Samoa. By ROBERT LOUIS STEVENSON. 6s.

Home Life of the Ancient Greeks, The. Translated by ALICE ZIMMERN. Illustrated. *Cheap Edition.* 5s.

Horse, The Book of the. By SAMUEL SIDNEY. With 17 Full-page Collotype Plates of Celebrated Horses of the Day, and numerous other Illustrations. Cloth, 15s.

Horses and Dogs. By O. EERELMAN. With Descriptive Text. Translated from the Dutch by CLARA BELL. With Photogravure Frontispiece, 12 Exquisite Collotypes, and several full page and other engravings in the text. 25s. net.

Houghton, Lord : The Life, Letters, and Friendships of Richard Monckton Milnes, First Lord Houghton. By Sir WEMYSS REID. In Two Vols., with Two Portraits. 32s.

Household, Cassell's Book of the. Complete in Four Vols. 5s. each. Four Vols. in Two, half-morocco, 25s.

Hygiene and Public Health. By B. ARTHUR WHITELEGGE, M.D. 7s. 6d.

Impregnable City, The. By MAX PEMBERTON. 6s.

Iron Pirate, The. By MAX PEMBERTON. Illustrated. 5s.

Island Nights' Entertainments. By R. L. STEVENSON. Illustrated. 6s.

Kennel Guide, The Practical. By Dr. GORDON STABLES. 1s.

Khiva, A Ride to. By Col. FRED BURNABY. *New Edition.* With Portrait and Seven Illustrations. 3s. 6d.

King George, In the Days of. By COL. PERCY GROVES. Illd. 1s. 6d.

Ladies' Physician, The. By a London Physician. *Cheap Edition, Revised and Enlarged.* 3s. 6d.

Lady Biddy Fane, The Admirable. By FRANK BARRETT. *New Edition.* With 12 Full-page Illustrations. 6s.

Lady's Dressing-room, The. Translated from the French of BARONESS STAFFE by LADY COLIN CAMPBELL. *Cheap Edition,* 2s. 6d.

Letters, The Highway of, and its Echoes of Famous Footsteps. By THOMAS ARCHER. Illustrated. *Cheap Edition,* 5s.

Letts's Diaries and other Time-saving Publications published exclusively by CASSELL & COMPANY. (*A list free on application.*) 6s.

'Lisbeth. A Novel. By LESLIE KEITH. 6s.

Little Minister, The. By J. M. BARRIE. *Illustrated Edition.* 6s.

Locomotive Engine, The Biography of a. By HENRY FRITH. 3s. 6d.

Loftus, Lord Augustus, The Diplomatic Reminiscences of. First and Second Series. Two Vols., each with Portrait, 32s. each Series.

London, Greater. By EDWARD WALFORD. Two Vols. With about 400 Illustrations. 9s. each.

London, Old and New. Six Vols., each containing about 200 Illustrations and Maps. Cloth, 9s. each.

London, Cassell's Guide to. With Numerous Illustrations. 6d.

London, The Queen's. With nearly 400 superb Views. 9s.

Lost on Du Corrig ; or, 'Twixt Earth and Ocean. By STANDISH O'GRADY. With 8 Full-page Illustrations. 5s.

Loveday: A Tale of a Stirring Time. By A. E. WICKHAM. Illustrated. 6s.

Manchester, Old and New. By WILLIAM ARTHUR SHAW, M.A. With Original Illustrations. Three Vols., 31s. 6d.

Medicine, Manuals for Students of. (*A List forwarded post free.*)

Modern Europe, A History of. By C. A. FYFFE, M.A. *Cheap Edition in One Volume,* 10s. 6d. Library Edition. Illustrated. 3 Vols., 7s. 6d. each.

Mrs. Cliff's Yacht. By FRANK STOCKTON. Illustrated. 6s.

Music, Illustrated History of. By EMIL NAUMANN. Edited by the Rev. Sir F. A. GORE OUSELEY, Bart. Illustrated. Two Vols. 31s. 6d.

National Library, Cassell's. In 214 Volumes. Paper covers, 3d.; cloth, 6d. (*A Complete List of the Volumes post free on application.*)

Natural History, Cassell's Concise. By E. PERCEVAL WRIGHT, M.A., M.D., F.L.S. With several Hundred Illustrations. 7s. 6d.

Natural History, Cassell's New. Edited by Prof. P. MARTIN DUNCAN, M.B., F.R.S., F.G.S. Complete in Six Vols. With about 2,000 Illustrations. Cloth, 9s. each.

Nature's Wonder Workers. By KATE R. LOVELL. Illustrated. 3s. 6d.

New Zealand, Pictorial. With Preface by Sir W. B. PERCEVAL, K.C.M.G. Illustrated. 6s.

Nursing for the Home and for the Hospital, A Handbook of. By CATHERINE J. WOOD. *Cheap Edition.* 1s. 6d.; cloth, 2s.

Nursing of Sick Children, A Handbook for the. By CATHERINE J. WOOD. 2s. 6d.

Oil Painting, A Manual of. By the Hon. JOHN COLLIER. 2s. 6d.

Old Maids and Young. By ELSA D'ESTERRE-KEELING. 6s.

Old Boy's Yarns, An. By HAROLD AVERY. With 8 Plates. 3s. 6d.

Our Own Country. Six Vols. With 1,200 Illustrations. 7s. 6d. each.

Painting, The English School of. *Cheap Edition.* 3s. 6d.

Painting, Practical Guides to. With Coloured Plates:—
MARINE PAINTING, 5s.; ANIMAL PAINTING, 5s.: CHINA PAINTING, 5s.; FIGURE PAINTING, 7s. 6d.; ELEMENTARY FLOWER PAINTING, 3s.; WATER-COLOUR PAINTING, 5s.; NEUTRAL TINT, 5s.; SEPIA, in Two Vols., 3s. each, or in One Vol., 5s.; FLOWERS, AND HOW TO PAINT THEM, 5s.

Paris, Old and New. Profusely Illustrated. In Two Vols.. 9s. each; or gilt edges, 10s. 6d. each.

Parliament, A Diary of the Home Rule, 1892-95. By H. W. LUCY. 10s. 6d.

Peoples of the World, The. In Six Vols. By Dr. ROBERT BROWN. Illustrated. 7s. 6d. each.

Photography for Amateurs. By T. C. HEPWORTH. *Enlarged and Revised Edition.* Illustrated. 1s.; or cloth, 1s. 6d.

Phrase and Fable, Dr. Brewer's Dictionary of. *Entirely New and Greatly Enlarged Edition.* 10s. 6d. Also in half morocco.

Picturesque America. Complete in Four Vols., with 48 Exquisite Steel Plates and about 800 Original Wood Engravings. £12 12s. the set. *Popular Edition*, Vols. I., II., & III., 18s. each.

Picturesque Australasia, Cassell's. With Upwards of 1,000 Illustrations. In Four Vols., 7s. 6d. each.   [the Set.

Picturesque Canada. With 600 Original Illustrations. Two Vols. £9 9s.

Picturesque Europe. Complete in Five Vols. Each containing 13 Exquisite Steel Plates, from Original Drawings, and nearly 200 Original Illustrations. Cloth, £21. POPULAR EDITION. In Five Vols., 18s. each

Picturesque Mediterranean, The. With Magnificent Original Illustrations by the leading Artists of the Day. Complete in Two Vols. £2 2s. each.

Pigeon Keeper, The Practical. By LEWIS WRIGHT. Illustrated. 3s. 6d.

Pigeons, Fulton's Book of. Edited by LEWIS WRIGHT. Revised, Enlarged and supplemented by the Rev. W. F. LUMLEY. With 50 Full-page Illustrations. *Popular Edition.* 10s. 6d. Original Edition, with 50 Coloured Plates and Numerous Wood-Engravings. 21s.

Planet, The Story of Our. By the Rev. Prof. BONNEY, F.R.S., etc. With Coloured Plates and Maps and about 100 Illustrations. *Cheap Edition.* 10s. 6d.

Pocket Library, Cassell's. Cloth, 1s. 4d. each.
A King's Diary. By PERCY WHITE. A White Baby. By JAMES WELSH. The Little Huguenot. By MAX PEMBERTON. A Whirl Asunder. By GERTRUDE ATHERTON. Lady Bonnie's Experiment. By TIGHE HOPKINS. The Paying Guest. By GEORGE GISSING.

Portrait Gallery, The Cabinet. Complete in Five Series, each containing 36 Cabinet Photographs of Eminent Men and Women. 15s. each.

Portrait Gallery, Cassell's Universal. Containing 240 Portraits of Celebrated Men and Women of the Day. Cloth, 6s.

Poultry Keeper, The Practical. By L. WRIGHT. Illustrated. 3s. 6d.

Poultry, The Book of. By LEWIS WRIGHT *Popular Edition.* 10s. 6d.

Poultry, The Illustrated Book of. By LEWIS WRIGHT. With Fifty Coloured Plates. *New and Revised Edition.* Cloth, gilt edges, 21s. Half-morocco *(Price on application).*

"Punch," The History of. By M. H. SPIELMANN. With nearly 170 Illustrations, Portraits, and Facsimiles. Cloth, 16s.; *Large Paper Edition,* £2 2s. net.

Puritan's Wife, A. By MAX PEMBERTON. Illustrated. 6s.

Q's Works, Uniform Edition of. 5s. each.

> Dead Man's Rock. The Splendid Spur. The Blue Pavilions. The Astonishing History of Troy Town. "I Saw Three Ships," and other Winter's Tales. Noughts and Crosses. The Delectable Duchy.

Queen Summer; or, The Tourney of the Lily and the Rose. With Forty Pages of Designs in Colours by WALTER CRANE. 6s.

Queen Victoria, The Life and Times of. By ROBERT WILSON. Complete in Two Vols. With numerous Illustrations. 9s. each.

Queen's London, The. Containing nearly 400 Exquisite Views of London and its Environs. Cloth, 9s.

Queen's Scarlet, The. By G. MANVILLE FENN. Illustrated. 3s. 6d.

Rabbit-Keeper, The Practical. By CUNICULUS. Illustrated. 3s. 6d.

Railways, Our. Their Origin, Development, Incident, and Romance. By JOHN PENDLETON. Illustrated. 2 Vols., 12s.

Railway Guides, Official Illustrated. With Illustrations, Maps, &c. Price 1s. each; or in cloth, 2s. each.

> LONDON AND NORTH WESTERN RAILWAY, GREAT WESTERN RAILWAY, MIDLAND RAILWAY, GREAT NORTHERN RAILWAY, GREAT EASTERN RAILWAY, LONDON AND SOUTH WESTERN RAILWAY, LONDON, BRIGHTON AND SOUTH COAST RAILWAY, SOUTH-EASTERN RAILWAY.

Railway Guides, Official Illustrated. Abridged and Popular Editions. Paper covers, 3d. each.

> GREAT EASTERN RAILWAY, LONDON AND NORTH WESTERN RAILWAY, LONDON AND SOUTH WESTERN RAILWAY, GREAT WESTERN RAILWAY, MIDLAND RAILWAY, GREAT NORTHERN RAILWAY, LONDON, BRIGHTON AND SOUTH COAST RAILWAY, SOUTH EASTERN RAILWAY.

Rivers of Great Britain: Descriptive, Historical, Pictorial.

> THE ROYAL RIVER: The Thames, from Source to Sea. 16s.
> RIVERS OF THE EAST COAST. *Popular Edition,* 16s.

Robinson Crusoe, Cassell's New Fine-Art Edition. 7s. 6d. *Cheap Edition,* 3s. 6d. or 5s.

Rogue's March, The. By E. W. HORNUNG. 6s.

Royal Academy Pictures, 1895. With upwards of 200 magnificent reproductions of Pictures in the Royal Academy of 1895. 7s. 6d.

Russo-Turkish War, Cassell's History of. With about 500 Illustrations. Two Vols., 9s. each. *New Edition,* Vol. I., 9s.

Sala, George Augustus, The Life and Adventures of. By Himself. *Library Edition,* in Two Vols., 32s. *Cheap Edition,* One Vol., 7s 6d.

Saturday Journal, Cassell's. Yearly Volume, cloth, 7s. 6d.

Science Series, The Century. Consisting of Biographies of Eminent Scientific Men of the present Century. Edited by Sir HENRY ROSCOE, D.C.L., F.R.S. Crown 8vo, 3s. 6d. each.

John Dalton and the Rise of Modern Chemistry. By Sir HENRY E. ROSCOE, F.R.S.
Major Rennell, F.R.S., and the Rise of English Geography. By CLEMENTS R. MARKHAM, C.B., F.R.S.
Justus Von Liebig: His Life and Work. By W. A. SHENSTONE, F.I.C.
The Herschels and Modern Astronomy. By MISS AGNES M. CLERKE.
Charles Lyell and Modern Geology. By Professor T. G. BONNEY, F.R.S.
J. Clerk Maxwell and Modern Physics. By R. T. GLAZEBROOK, F.R.S.
Sir Humphry Davy, Poet and Philosopher. By T. E. THORPE, F.R.S.
Charles Darwin and the Theory of Natural Selection. By EDWARD B POULTON, M.A., F.R.S.

Science for All. Edited by Dr. ROBERT BROWN. Five Vols. 9s. each.

Scotland, Picturesque and Traditional. By G. E. EYRE-TODD. 6s.

Sea, The Story of the. An Entirely New and Original Work. Edited by Q. Illustrated. In Two Vols., 9s. each.

Sea-Wolves, The. By MAX PEMBERTON. Illustrated. 6s.

Sentimental Tommy. By J. M. BARRIE. 6s.

Shaftesbury, The Seventh Earl of, K.G., The Life and Work of. By EDWIN HODDER. *Cheap Edition.* 3s. 6d.

Shakespeare, The Plays of. Edited by Professor HENRY MORLEY. Complete in Thirteen Vols., cloth, 21s.; also 39 Vols., cloth, in box, 21s.; half-morocco, cloth sides, 42s.

Shakespeare, Cassell's Quarto Edition. Containing about 600 Illustrations by H. C. SELOUS. Complete in Three Vols., cloth gilt, £3 3s.

Shakespeare, The England of. *New Edition.* By E. GOADBY. With Full-page Illustrations. 2s. 6d.

Shakspere's Works. *Édition de Luxe.*
"King Henry VIII." Illustrated by SIR JAMES LINTON, P.R.I. (*Price on application.*)
"Othello." Illustrated by FRANK DICKSEE, R.A. £3 10s.
"King Henry IV." Illustrated by EDUARD GRÜTZNER. £3 10s.
"As You Like It." Illustrated by EMILE BAYARD. £3 10s.

Shakspere, The Leopold. With 400 Illustrations. *Cheap Edition.* 3s. 6d. Cloth gilt, gilt edges, 5s.; Roxburgh, 7s. 6d.

Shakspere, The Royal. With Steel Plates and Wood Engravings. Three Vols. 15s. each.

Sketches, The Art of Making and Using. From the French of G. FRAIPONT. By CLARA BELL. With 50 Illustrations. 2s. 6d.

Social England. A Record of the Progress of the People. By various writers. Edited by H. D. TRAILL, D.C.L. Vols. I., II., & III., 15s. each. Vols. IV. & V., 17s. each. Vol. VI., 18s.

Songs for Soldiers and Sailors. By JOHN FARMER. 5s. Words only, 6d.

Sports and Pastimes, Cassell's Complete Book of. *Cheap Edition.* With more than 900 Illustrations. Medium 8vo, 992 pages, cloth, 3s. 6d.

Star-Land. By Sir R. S. BALL, LL.D., &c. Illustrated. 6s.

Story of Francis Cludde, The. By STANLEY J. WEYMAN. 6s.

Story of my Life, The. By SIR RICHARD TEMPLE. Two Vols. 21s.

Sun, The. By Sir ROBERT STAWELL BALL, LL.D., F.R.S., F.R.A.S. With Eight Coloured Plates and other Illustrations. 21s.

The "Treasure Island" Series. *Illustrated Edition.* 3s. 6d. each.

Treasure Island. By ROBERT LOUIS STEVENSON.	**The Black Arrow.** By ROBERT LOUIS STEVENSON.
The Master of Ballantrae. By ROBERT LOUIS STEVENSON.	**King Solomon's Mines.** By H. RIDER HAGGARD.

Selections from Cassell & Company's Publications.

Things I have Seen and People I have Known. By G. A. SALA. With Portrait and Autograph. 2 Vols. 21s.

Tidal Thames, The. By GRANT ALLEN. With India Proof Impressions of Twenty magnificent Full-page Photogravure Plates, and with many other Illustrations in the Text after Original Drawings by W. L. WYLLIE, A.R.A. *New Edition*, cloth, 42s. net. A'so in Half morocco.

To the Death. By R. D. CHETWODE. With Four Plates, 5s.

Treatment, The Year-Book of, for 1897. A Critical Review for Practitioners of Medicine and Surgery. *Thirteenth Year of Issue.* 7s. 6d.

Trees, Familiar. By G. S. BOULGER, F.L.S. Two Series. With 40 Coloured Plates in each. (*Price on application.*)

Tuxter's Little Maid. By G. B. BURGIN. 6s.

Uncle Tom's Cabin. By HARRIET BEECHER STOWE. With upwards of 100 Original Illustrations. *Fine Art Memorial Edition*, 7s. 6d.

"Unicode": the Universal Telegraphic Phrase Book. *Desk or Pocket Edition.* 2s. 6d.

United States, Cassell's History of the. By EDMUND OLLIER. With 600 Illustrations. Three Vols. 9s. each.

Universal History, Cassell's Illustrated. Four Vols. 9s. each.

Vision of Saints, A. By Sir LEWIS MORRIS. With 20 Full-page Illustrations. Crown 4to, cloth, 10s. 6d. *Non-illustrated Edition*, 6s.

Wandering Heath. Short Stories. By Q. 6s.

War and Peace, Memories and Studies of. By ARCHIBALD FORBES. *Original Edition*, 16s. *Cheap Edition*, 6s.

Westminster Abbey, Annals of. By E. T. BRADLEY (Mrs. A. MURRAY SMITH). Illustrated. With a Preface by the Dean of Westminster. 63s.

What Cheer! By W. CLARK RUSSELL. 6s.

White Shield, The. By BERTRAM MITFORD. 6s.

Wild Birds, Familiar. By W. SWAYSLAND. Four Series. With 40 Coloured Plates in each. (Sold in sets only; price on application.)

Wild Flowers, Familiar. By F. E. HULME, F.L.S., F.S.A. Five Series. With 40 Coloured Plates in each. (In sets only; price on application.)

Windsor Castle, The Governor's Guide to. By the Most Noble the MARQUIS OF LORNE, K.T. Illustrated. 1s.; cloth, 2s.

Wit and Humour, Cassell's New World of. With New Pictures and New Text. In Two Vols., 6s. each.

With Claymore and Bayonet. By Col. PERCY GROVES. Illd. 5s.

Work. The Illustrated Weekly Journal for Mechanics. Half-yearly. Vols., 4s. each.

"Work" Handbooks. Practical Manuals prepared *under the direction of* PAUL N. HASLUCK, Editor of *Work*. Illustrated. 1s. each.

World of Wonders. Illustrated. *Cheap Edition*, Vol. I., 4s. 6d.

Wrecker, The. By R. L. STEVENSON and L. OSBOURNE. Illustrated. 6s.

ILLUSTRATED MAGAZINES.

The Quiver. Monthly, 6d.

Cassell's Family Magazine. Monthly, 6d.

"Little Folks" Magazine. Monthly, 6d.

The Magazine of Art. Monthly, 1s. 4d.

"Chums." Illustrated Paper for Boys. Weekly, 1d.; Monthly, 6d.

Cassell's Saturday Journal. Weekly, 1d.; Monthly, 6d.

Work. Weekly, 1d.; Monthly, 6d.

Building World. The New Practical Journal on Building and Building Trades. Weekly, 1d.; Monthly, 6d.

Cottage Gardening. Weekly, ½d.; Monthly, 3d.

CASSELL & COMPANY, LIMITED, *Ludgate Hill, London.*

Bibles and Religious Works.

Bible Biographies. Illustrated. 1s. 6d. each.
The Story of Moses and Joshua. By the Rev. J. TELFORD.
The Story of the Judges. By the Rev. J. WYCLIFFE GEDGE.
The Story of Samuel and Saul. By the Rev. D. C. TOVEY.
The Story of David. By the Rev. J. WILD.
The Story of Joseph. Its Lessons for To-Day. By the Rev. GEORGE BAINTON.

The Story of Jesus. In Verse. By J. R. MACDUFF, D.D.

Bible, Cassell's Illustrated Family. With 900 Illustrations. Leather, gilt edges, £2 10s.

Bible Educator, The. Edited by the Very Rev. Dean PLUMPTRE, D.D With Illustrations, Maps, &c. Four Vols., cloth, 6s. each.

Bible Dictionary, Cassell's Concise Illustrated. By the Rev. ROBERT HUNTER, LL.D. *Illustrated.* 7s. 6d.

Bible Student in the British Museum, The. By the Rev. J. G. KITCHIN, M.A. *New and Revised Edition.* 1s. 4d.

Bunyan, Cassell's Illustrated. With 200 Original Illustrations. *Cheap Edition.* 7s. 6d.

Bunyan's Pilgrim's Progress. Illustrated throughout. Cloth, 3s. 6d.; cloth gilt, gilt edges, 5s.

Child's Bible, The. With 200 Illustrations. *150th Thousand.* 7s. 6d.

Child's Life of Christ, The. With 200 Illustrations. 7s. 6d.

Conquests of the Cross. Illustrated. In 3 Vols. 9s. each.

Doré Bible With 238 Illustrations by GUSTAVE DORÉ. Small folio, best morocco, gilt edges, £15. *Popular Edition.* With 200 Illustrations. 15s.

Early Days of Christianity, The. By the Very Rev. Dean FARRAR, D.D., F.R.S. LIBRARY EDITION. Two Vols., 24s.; morocco, £2 2s. POPULAR EDITION. Complete in One Volume, cloth, 6s.; cloth, gilt edges. 7s. 6d.; Persian morocco, 10s. 6d.; tree-calf, 15s.

Family Prayer-Book, The. Edited by Rev. Canon GARBETT, M.A., and Rev. S. MARTIN. With Full-page Illustrations. 7s. 6d.

Gleanings after Harvest. Studies and Sketches by the Rev. JOHN R. VERNON, M.A. Illustrated. 6s.

"Graven in the Rock." By the Rev. Dr. SAMUEL KINNS, F.R.A.S. Illustrated. *Library Edition.* Two Vols., 15s.

"Heart Chords." A Series of Works by Eminent Divines. 1s. each.

MY COMFORT IN SORROW. By HUGH MACMILLAN, D.D.
MY BIBLE. By the Right Rev. W. BOYD CARPENTER, Bishop of Ripon.
MY FATHER. By the Right Rev. ASHTON OXENDEN, late Bishop of Montreal.
MY WORK FOR GOD. By the Right Rev. Bishop COTTERILL.
MY OBJECT IN LIFE. By the Very Rev. Dean FARRAR, D.D.
MY ASPIRATIONS. By the Rev. G. MATHESON, D.D.
MY EMOTIONAL LIFE. By the Rev. Preb. CHADWICK, D.D.

MY BODY. By the Rev. Prof. W. G. BLAIKIE, D.D.
MY GROWTH IN DIVINE LIFE. By the Rev. Preb. REYNOLDS, M.A.
MY SOUL. By the Rev. P. B. POWER, M.A.
MY HEREAFTER. By the Very Rev. Dean BICKERSTETH.
MY WALK WITH GOD. By the Very Rev. Dean MONTGOMERY.
MY AIDS TO THE DIVINE LIFE. By the Very Rev. Dean BOYLE.
MY SOURCES OF STRENGTH. By the Rev. E. E. JENKINS, M.A., Secretary of Wesleyan Missionary Society.

Helps to Belief. A Series of Helpful Manuals on the Religious Difficulties of the Day. Edited by the Rev. TEIGNMOUTH SHORE, M.A., Canon of Worcester. Cloth, 1s. each.

CREATION. By Harvey Goodwin, D.D., late Bishop of Carlisle.
THE DIVINITY OF OUR LORD. By the Lord Bishop of Derry.
MIRACLES. By the Rev. Brownlow Maitland, M.A.

PRAYER. By the Rev. Canon Shore, M.A.
THE ATONEMENT. By William Connor Magee, D.D., Late Archbishop of York.

Holy Land and the Bible, The. By the Rev. C. Gеikie, D.D., LL.D. (Edin.). *Cheap Edition,* with 24 Collotype Plates, 12s. 6d.

Life of Christ, The. By the Very Rev. Dean Farrar, D.D., F.R.S. Cheap Edition. With 16 Full-page Plates. Cloth gilt, 3s. 6d. Library Edition. Two Vols. Cloth, 24s.; morocco, 42s. Illustrated Edition. Cloth, 7s. 6d.; cloth, full gilt, gilt edges, 10s. 6d. Popular Edition (*Revised and Enlarged*), 8vo, cloth, gilt edges, 7s. 6d.; Persian morocco, gilt edges, 10s. 6d.; tree-calf, 15s.

Moses and Geology; or, The Harmony of the Bible with Science. By the Rev. Samuel Kinns, Ph.D., F.R.A.S. Illustrated *Library Edition, Revised to Date.* 10s. 6d.

My Last Will and Testament. By Hyacinthe Loyson (Père Hyacinthe). Translated by Fabian Ware. 1s.; cloth, 1s. 6d.

New Light on the Bible and the Holy Land. By B. T. A. Evetts, M.A. Illustrated. 7s. 6d

New Testament Commentary for English Readers, The. Edited by Bishop Ellicott. In Three Volumes. 21s. each. Vol. I.—The Four Gospels. Vol. II.—The Acts, Romans, Corinthians, Galatians. Vol. III.—The remaining Books of the New Testament.

New Testament Commentary. Edited by Bishop Ellicott. Handy Volume Edition. St. Matthew, 3s. 6d. St. Mark, 3s. St. Luke, 3s. 6d. St. John, 3s. 6d. The Acts of the Apostles, 3s. 6d. Romans, 2s. 6d. Corinthians I. and II., 3s. Galatians, Ephesians, and Philippians, 3s. Colossians, Thessalonians, and Timothy, 3s. Titus, Philemon, Hebrews, and James, 3s. Peter, Jude, and John, 3s. The Revelation, 3s. An Introduction to the New Testament, 3s. 6d.

Old Testament Commentary for English Readers, The. Edited by Bishop Ellicott. Complete in Five Vols. 21s. each. Vol. I.—Genesis to Numbers. Vol. II. — Deuteronomy to Samuel II. Vol. III. — Kings I. to Esther. Vol. IV.—Job to Isaiah. Vol. V.—Jeremiah to Malachi.

Old Testament Commentary. Edited by Bishop Ellicott. Handy Volume Edition. Genesis, 3s. 6d. Exodus, 3s. Leviticus, 3s. Numbers, 2s. 6d. Deuteronomy, 2s. 6d.

Plain Introductions to the Books of the Old Testament. Edited by Bishop Ellicott. 3s. 6d.

Plain Introductions to the Books of the New Testament. Edited by Bishop Ellicott. 3s. 6d.

Protestantism, The History of. By the Rev. J. A. Wylie, LL.D. Containing upwards of 600 Original Illustrations. Three Vols. 9s. each.

Quiver Yearly Volume, The. With about 600 Original Illustrations. 7s. 6d.

Religion, The Dictionary of. By the Rev. W. Benham, B.D. *Cheap Edition.* 10s. 6d.

St. George for England; and other Sermons preached to Children. By the Rev. T. Teignmouth Shore, M.A., Canon of Worcester. 5s.

St. Paul, The Life and Work of. By the Very Rev. Dean Farrar, D.D., F.R.S. Library Edition. Two Vols., cloth, 24s.; calf, 42s. Illustrated Edition, complete in One Volume, with about 300 Illustrations, £1 1s.; morocco, £2 2s. Popular Edition. One Volume, 8vo, cloth, 6s.; Persian morocco, 10s. 6d.; tree-calf, 15s.

Shall We Know One Another in Heaven? By the Rt. Rev. J. C. Ryle, D.D., Bishop of Liverpool. *Cheap Edition.* Paper covers, 6d.

Searchings in the Silence. By Rev. George Matheson, D.D. 3s. 6d.

"Sunday," Its Origin, History, and Present Obligation. By the Ven. Archdeacon Hessey, D.C.L. *Fifth Edition.* 7s. 6d.

Twilight of Life, The. Words of Counsel and Comfort for the Aged. By the Rev. John Ellerton, M.A. 1s. 6d.

Educational Works and Students' Manuals.

Agricultural Text-Books, Cassell's. (The "Downton" Series.) Edited by JOHN WRIGHTSON, Professor of Agriculture. Fully Illustrated, 2s. 6d. each.—Farm Crops. By Prof. WRIGHTSON.—Soils and Manures. By J. M. H. MUNRO, D.Sc. (London), F.I.C., F.C.S. —Live Stock. By Prof. WRIGHTSON.

Alphabet, Cassell's Pictorial. 3s. 6d.

Arithmetics, Cassell's "Belle Sauvage." By GEORGE RICKS, B.Sc. Lond. With Test Cards. (*List on application.*)

Atlas, Cassell's Popular. Containing 24 Coloured Maps. 1s. 6d.

Book-Keeping. By THEODORE JONES. For Schools, 2s.; cloth, 3s. For the Million, 2s.; cloth, 3s. Books for Jones's System, 2s.

British Empire Map of the World. New Map for Schools and Institutes. By G. R. PARKIN and J. G. BARTHOLOMEW, F.R.G.S. 25s.

Chemistry, The Public School. By J. H. ANDERSON, M.A. 2s. 6d.

Cookery for Schools. By LIZZIE HERITAGE. 6d.

Dulce Domum. Rhymes and Songs for Children. Edited by JOHN FARMER, Editor of "Gaudeamus," &c. Old Notation and Words, 5s. N.B.—The words of the Songs in "Dulce Domum" (with the Airs both in Tonic Sol-fa and Old Notation) can be had in Two Parts, 6d. each.

Euclid, Cassell's. Edited by Prof. WALLACE, M.A. 1s.

Euclid, The First Four Books of. *New Edition.* In paper, 6d.; cloth, 9d.

Experimental Geometry. By PAUL BERT. Illustrated. 1s. 6d.

French, Cassell's Lessons in. *New and Revised Edition.* Parts I. and II., each 2s. 6d.; complete, 4s. 6d. Key, 1s. 6d.

French-English and English-French Dictionary. *Entirely New and Enlarged Edition.* Cloth, 3s. 6d.; half morocco, 5s.

French Reader, Cassell's Public School. By G. S. CONRAD. 2s. 6d.

Gaudeamus. Songs for Colleges and Schools. Edited by JOHN FARMER. 5s. Words only, paper covers, 6d.; cloth, 9d.

German Dictionary, Cassell's New (German-English, English-German). *Cheap Edition.* Cloth 3s. 6d.; half morocco, 5s.

Hand and Eye Training. By G. RICKS, B.Sc. 2 Vols., with 16 Coloured Plates in each Vol. Cr. 4to, 6s. each. Cards for Class Use, 5 sets, 1s. each.

Hand and Eye Training. By GEORGE RICKS, B.Sc., and JOSEPH VAUGHAN. Illustrated. Vol. I. Designing with Coloured Papers; Vol. II. Cardboard Work, 2s. each. Vol. III. Colour Work and Design, 3s.

Historical Cartoons, Cassell's Coloured. Size 45 in. × 35 in., 2s. each. Mounted on canvas and varnished, with rollers, 5s. each.

Italian Lessons, with Exercises, Cassell's. Cloth, 3s. 6d.

Latin Dictionary, Cassell's New. (Latin-English and English-Latin.) Revised by J. R. V. MARCHANT, M.A., and J. F. CHARLES, B.A. Cloth, 3s. 6d.; half morocco, 5s.

Latin Primer, The First. By Prof. POSTGATE. 1s.

Latin Primer, The New. By Prof. J. P. POSTGATE. Crown 8vo, 2s. 6d.

Latin Prose for Lower Forms. By M. A. BAYFIELD, M.A. 2s. 6d.

Laws of Every-Day Life. By H. O. ARNOLD-FORSTER, M.P. 1s. 6d. *Special Edition* on Green Paper for Persons with Weak Eyesight. 2s.

Lessons in Our Laws; or, Talks at Broadacre Farm. By H. F. LESTER, B.A. Parts I. and II., 1s. 6d. each.

Little Folks' History of England. Illustrated. 1s. 6d.

Making of the Home, The. By Mrs. SAMUEL A. BARNETT. 1s. 6d.

Marlborough Books:—Arithmetic Examples, 3s. French Exercises, 3s. 6d. French Grammar, 2s. German Grammar, 3s. 6d.

Mechanics and Machine Design, Numerical Examples in Practical. By R. G. BLAINE, M.E. *New Edition, Revised and Enlarged.* With 79 Illustrations. Cloth, 2s. 6d.

Mechanics for Young Beginners, A First Book of. By the Rev. J. G. EASTON, M.A. *Cheap Edition.* 2s. 6d.

Natural History Coloured Wall Sheets, Cassell's New. 16 Subjects. Size 39 by 31 in. Mounted on rollers and varnished. 3s. each.
Object Lessons from Nature. By Prof. L. C. MIALL, F.L.S. Fully Illustrated. *New and Enlarged Edition.* Two Vols., 1s. 6d. each.
Physiology for Schools. By A. T. SCHOFIELD, M.D., M.R.C.S., &c. Illustrated. Cloth, 1s. 9d.; Three Parts, paper covers, 5d. each; or cloth limp, 6d. each.
Poetry Readers, Cassell's New. Illustrated. 12 Books, 1d. each; or complete in one Vol., cloth, 1s. 6d.
Popular Educator, Cassell's NEW. With Revised Text, New Maps, New Coloured Plates, New Type, &c. In Vols., 5s. each; or in Four Vols., half-morocco, 50s. the set.
Readers, Cassell's "Belle Sauvage." An entirely New Series. Fully Illustrated. Strongly bound in cloth. (*List on application.*)
Readers, Cassell's "Higher Class." (*List on application.*)
Readers, Cassell's Readable. Illustrated. (*List on application.*)
Readers for Infant Schools, Coloured. Three Books. 4d. each.
Reader, The Citizen. By H. O. ARNOLD-FORSTER, M.P. Illustrated. 1s. 6d. Also a *Scottish Edition,* cloth, 1s. 6d.
Reader, The Temperance. By Rev. J. DENNIS HIRD. Crown 8vo, 1s. 6d.
Readers, Geographical, Cassell's New. With numerous Illustrations. (*List on application.*)
Readers, The "Modern School" Geographical. (*List on application.*)
Readers, The "Modern School." Illustrated. (*List on application.*)
Reckoning, Howard's Art of. By C. FRUSHER HOWARD. Paper covers, 1s.; cloth, 2s. *New Edition,* 5s.
Round the Empire. By G. R. PARKIN. Fully Illustrated. 1s. 6d.
Science Applied to Work. By J. A. BOWER. 1s.
Science of Everyday Life By J. A. BOWER. Illustrated. 1s.
Shade from Models, Common Objects, and Casts of Ornament, How to. By W. E. SPARKES. With 25 Plates by the Author. 3s.
Shakspere's Plays for School Use. 9 Books. Illustrated. 6d. each.
Spelling, A Complete Manual of. By J. D. MORELL, LL.D. 1s.
Technical Manuals, Cassell's. Illustrated throughout:—
Handrailing and Staircasing, 3s. 6d.—Bricklayers, Drawing for, 3s.—Building Construction, 2s. — Cabinet-Makers, Drawing for, 3s. — Carpenters and Joiners, Drawing for, 3s. 6d.—Gothic Stonework, 3s.—Linear Drawing and Practical Geometry, 2s.—Linear Drawing and Projection. The Two Vols. in One, 3s. 6d.—Machinists and Engineers, Drawing for, 4s. 6d.—Model Drawing, 3s.—Orthographical and Isometrical Projection, 2s.—Practical Perspective, 3s.—Stonemasons, Drawing for, 3s.—Applied Mechanics, by Sir R. S. BALL, LL.D., 2s.—Systematic Drawing and Shading, 2s.
Technical Educator, Cassell's New. With Coloured Plates and Engravings. Complete in Six Volumes, 5s. each.
Technology, Manuals of. Edited by Prof. AYRTON, F.R.S., and RICHARD WORMELL, D.Sc., M.A. Illustrated throughout:—
The Dyeing of Textile Fabrics, by Prof. Hummel, 5s.—Watch and Clock Making, by D. Glasgow, Vice-President of the British Horological Institute, 4s. 6d.—Steel and Iron, by Prof. W. H. Greenwood, F.C.S., M.I.C.E., &c., 5s.—Spinning Woollen and Worsted, by W. S. B. McLaren, M.P., 4s. 6d.—Design in Textile Fabrics, by T. R. Ashenhurst, 4s. 6d.—Practical Mechanics, by Prof. Perry, M.E., 3s. 6d.—Cutting Tools Worked by Hand and Machine, by Prof. Smith, 3s. 6d.
Things New and Old; or, Stories from English History. By H. O. ARNOLD-FORSTER, M.P. Fully Illustrated, and strongly bound in Cloth. Standards I. & II., 9d. each; Standard III., 1s.; Standard IV., 1s. 3d.; Standards V. & VI., 1s. 6d. each; Standard VII., 1s. 8d.
This World of Ours. By H. O. ARNOLD-FORSTER, M.P. Illustrated. 3s. 6d.

Books for Young People.

"**Little Folks**" **Half-Yearly Volume.** Containing 480 4to pages, with Pictures on nearly every page, together with Six Full-page Coloured Plates, and numerous other Illustrations in Colour. Boards, 3s. 6d. ; cloth gilt, gilt edges, 5s. each.

Bo-Peep. A Book for the Little Ones. With Original Stories and Verses. With 8 Coloured Plates, and numerous other Illustrations printed in Colour. Yearly Volume. Boards, 2s. 6d. ; cloth, 3s. 6d.

Beneath the Banner. Being Narratives of Noble Lives and Brave Deeds. By F. J. CROSS. Illustrated. Limp cloth, 1s. Cloth gilt, 2s.

Good Morning! Good Night! By F. J. CROSS. Illustrated. Limp cloth, 1s., or cloth boards, gilt lettered, 2s.

Five Stars in a Little Pool. By EDITH CARRINGTON. Illustrated. 3s. 6d.

Merry Girls of England. By L. T. MEADE. 3s. 6d.

Beyond the Blue Mountains. By L. T. MEADE. 5s.

The Peep of Day. *Cassell's Illustrated Edition.* 2s. 6d.

A Book of Merry Tales. By MAGGIE BROWNE, "SHEILA," ISABEL WILSON, and C. L. MATÉAUX. Illustrated. 3s. 6d.

A Sunday Story-Book. By MAGGIE BROWNE, SAM BROWNE, and AUNT ETHEL. Illustrated. 3s. 6d.

A Bundle of Tales. By MAGGIE BROWNE (Author of "Wanted—a King," &c.), SAM BROWNE, and AUNT ETHEL. 3s. 6d.

Pleasant Work for Busy Fingers. By MAGGIE BROWNE. Illustrated *Cheap Edition.* 2s. 6d.

Born a King. By FRANCES and MARY ARNOLD-FORSTER. (The Life of Alfonso XIII., the Boy King of Spain.) Illustrated. 1s.

Cassell's Pictorial Scrap Book. In 24 Books, 6d. each.

Schoolroom and Home Theatricals. By ARTHUR WAUGH. Illustrated. *New Edition.* Cloth, 1s. 6d.

Magic at Home. By Prof. HOFFMAN. Illustrated. Cloth gilt, 3s. 6d.

Little Mother Bunch. By Mrs. MOLESWORTH. Illustrated. *New Edition.* Cloth. 2s. 6d.

Heroes of Every-day Life. By LAURA LANE. With about 20 Full-page Illustrations. Cloth. 2s. 6d.

Books for Young People. Illustrated. 3s. 6d. each.

The Champion of Odin; or, Viking Life in the Days of Old. By J. Fred. Hodgetts.

Bound by a Spell; or, The Hunted Witch of the Forest. By the Hon. Mrs. Greene.

Under Bayard's Banner. By Henry Frith.

Told Out of School. By A. J. Daniels.

*Red Rose and Tiger Lily.** By L. T. Meade.

The Romance of Invention. By James Burnley.

*Bashful Fifteen.** By L. T. Meade.

* **The White House at Inch Gow.** By Mrs. Pitt.

*A Sweet Girl Graduate.** By L. T. Meade.

The King's Command: A Story for Girls. By Maggie Symington.

*The Palace Beautiful.** By L. T. Meade.

*Polly: A New-Fashioned Girl.** By L. T. Meade.

"**Follow My Leader.**" By Talbot Baines Reed.

*A World of Girls: The Story of a School.** By L. T. Meade.

Lost among White Africans. By David Ker.

For Fortune and Glory: A Story of the Soudan War. By Lewis Hough.

Bob Lovell's Career. By Edward S. Ellis.

*Also procurable in superior binding, 5s. each.

"Peeps Abroad" Library. *Cheap Editions.* Gilt edges, 2s. 6d. each.

Rambles Round London. By C. L. Matéaux. Illustrated.

Around and About Old England. By C. L. Matéaux. Illustrated.

Paws and Claws. By one of the Authors of "Poems written for a Child." Illustrated.

Decisive Events in History. By Thomas Archer. With Original Illustrations.

The True Robinson Crusoes. Cloth gilt.

Peeps Abroad for Folks at Home. Illustrated throughout.

Wild Adventures in Wild Places. By Dr. Gordon Stables, R.N. Illustrated.

Modern Explorers. By Thomas Frost. Illustrated. *New and Cheaper Edition.*

Early Explorers. By Thomas Frost.

Home Chat with our Young Folks. Illustrated throughout.

Jungle, Peak, and Plain. Illustrated throughout.

The "Cross and Crown" Series. Illustrated. 2s. 6d. each.

Freedom's Sword: A Story of the Days of Wallace and Bruce. By Annie S. Swan.

Strong to Suffer: A Story of the Jews. By E. Wynne.

Heroes of the Indian Empire; or, Stories of Valour and Victory. By Ernest Foster.

In Letters of Flame: A Story of the Waldenses. By C. L. Matéaux.

Through Trial to Triumph. By Madeline B. Hunt.

By Fire and Sword: A Story of the Huguenots. By Thomas Archer.

Adam Hepburn's Vow: A Tale of Kirk and Covenant. By Annie S. Swan.

No. XIII.; or, The Story of the Lost Vestal. A Tale of Early Christian Days. By Emma Marshall.

"Golden Mottoes" Series, The. Each Book containing 208 pages, with Four full-page Original Illustrations. Crown 8vo, cloth gilt, 2s. each.

"Nil Desperandum." By the Rev. F. Langbridge, M.A.

"Bear and Forbear." By Sarah Pitt.

"Foremost if I Can." By Helen Atteridge.

"Honour is my Guide." By Jeanie Hering (Mrs. Adams-Acton).

"Aim at a Sure End." By Emily Searchfield.

"He Conquers who Endures." By the Author of "May Cunningham's Trial," &c.

"Wanted—a King" Series. Illustrated. 2s. 6d. each.

Great Grandmamma. By Georgina M. Synge.

Robin's Ride. By Ellinor Davenport Adams.

Wanted—a King; or, How Merle set the Nursery Rhymes to Rights. By Maggie Browne.

Fairy Tales in Other Lands. By Julia Goddard.

Cassell's Picture Story Books. Each containing about Sixty Pages of Pictures and Stories, &c. 6d. each.

Little Talks.
Bright Stars.
Nursery Toys.
Pet's Posy.
Tiny Tales.

Daisy's Story Book.
Dot's Story Book.
A Nest of Stories.
Good-Night Stories.
Chats for Small Chatterers.

Auntie's Stories.
Birdie's Story Book.
Little Chimes.
A Sheaf of Tales.
Dewdrop Stories.

Illustrated Books for the Little Ones. Containing interesting Stories. All Illustrated. 1s. each; cloth gilt, 1s. 6d.

Bright Tales & Funny Pictures.
Merry Little Tales.
Little Tales for Little People.
Little People and Their Pets.
Tales Told for Sunday.
Sunday Stories for Small People.
Stories and Pictures for Sunday.
Bible Pictures for Boys and Girls.
Firelight Stories.
Sunlight and Shade.
Rub-a-Dub Tales.
Fine Feathers and Fluffy Fur.
Scrambles and Scrapes.
Tittle Tattle Tales.

Up and Down the Garden.
All Sorts of Adventures.
Our Sunday Stories.
Our Holiday Hours.
Indoors and Out.
Some Farm Friends.
Wandering Ways.
Dumb Friends.
Those Golden Sands.
Little Mothers & their Children.
Our Pretty Pets.
Our Schoolday Hours.
Creatures Tame.
Creatures Wild.

Cassell's Shilling Story Books. All Illustrated, and containing Interesting Stories.

Bunty and the Boys.
The Heir of Elmdale.
The Mystery at Shoncliff School.
Claimed at Last, & Roy's Reward.
Thorns and Tangles.
The Cuckoo in the Robin's Nest.
John's Mistake. [Pitchers.
The History of Five Little Diamonds in the Sand.
Surly Bob.

The Giant's Cradle.
Shag and Doll.
Aunt Lucia's Locket.
The Magic Mirror.
The Cost of Revenge.
Clever Frank.
Among the Redskins.
The Ferryman of Brill.
Harry Maxwell.
A Banished Monarch.
Seventeen Cats.

The World's Workers. A Series of New and Original Volumes. With Portraits printed on a tint as Frontispiece. 1s. each.

John Cassell. By G. Holden Pike.
Charles Haddon Spurgeon. By G. Holden Pike.
Dr. Arnold of Rugby. By Rose E. Selfe.
The Earl of Shaftesbury. By Henry Frith.
Sarah Robinson, Agnes Weston, and Mrs. Meredith. By E. M. Tomkinson.
Thomas A. Edison and Samuel F. B. Morse. By Dr. Denslow and J. Marsh Parker.
Mrs. Somerville and Mary Carpenter. By Phyllis Browne.
General Gordon. By the Rev. S. A. Swaine.
Charles Dickens. By his Eldest Daughter.
Sir Titus Salt and George Moore. By J. Burnley.

Florence Nightingale, Catherine Marsh, Frances Ridley Havergal, Mrs. Ranyard ("L. N. R."). By Lizzie Alldridge.
Dr. Guthrie, Father Mathew, Elihu Burritt, George Livesey. By John W. Kirton, LL.D.
Sir Henry Havelock and Colin Campbell Lord Clyde. By E. C. Phillips.
Abraham Lincoln. By Ernest Foster.
George Müller and Andrew Reed. By E. R. Pitman.
Richard Cobden. By R. Gowing.
Benjamin Franklin. By E. M. Tomkinson.
Handel. By Eliza Clarke. [Swaine.
Turner the Artist. By the Rev. S. A.
George and Robert Stephenson. By C. L. Matéaux.
David Livingstone. By Robert Smiles.

⁎ *The above Works can also be had Three in One Vol., cloth, gilt edges, 3s.*

Library of Wonders. Illustrated Gift-books for Boys. Paper, 1s.; cloth, 1s. 6d.

Wonderful Balloon Ascents.
Wonderful Adventures.
Wonderful Escapes.

Wonders of Animal Instinct.
Wonders of Bodily Strength and Skill.

Cassell's Eighteenpenny Story Books. Illustrated.

Wee Willie Winkie.
Ups and Downs of a Donkey's Life.
Three Wee Ulster Lassies.
Up the Ladder.
Dick's Hero; and other Stories.
The Chip Boy.
Raggles, Baggles, and the Emperor.
Roses from Thorns.

Faith's Father.
By Land and Sea.
The Young Berringtons.
Jeff and Leff.
Tom Morris's Error.
Worth more than Gold.
"Through Flood—Through Fire"; and other Stories.
The Girl with the Golden Locks.
Stories of the Olden Time.

Gift Books for Young People. By Popular Authors. With Four Original Illustrations in each. Cloth gilt, 1s. 6d. each.

The Boy Hunters of Kentucky. By Edward S. Ellis.
Red Feather: a Tale of the American Frontier. By Edward S. Ellis.
Seeking a City.
Rhoda's Reward; or, "If Wishes were Horses."
Jack Marston's Anchor.
Frank's Life-Battle; or, The Three Friends.
Fritters. By Sarah Pitt.
The Two Hardcastles. By Madeline Bonavia Hunt.

Major Monk's Motto. By the Rev. F. Langbridge.
Trixy. By Maggie Symington.
Rags and Rainbows: A Story of Thanksgiving.
Uncle William's Charges; or, The Broken Trust.
Pretty Pink's Purpose; or, The Little Street Merchants.
Tim Thomson's Trial. By George Weatherly.
Ursula's Stumbling-Block. By Julia Goddard.
Ruth's Life-Work. By the Rev. Joseph Johnson.

Selections from Cassell & Company's Publications.

Cassell's Two-Shilling Story Books. Illustrated.

Margaret's Enemy.
Stories of the Tower.
Mr. Burke's Nieces.
May Cunningham's Trial.
The Top of the Ladder : How to Reach it.
Little Flotsam.
Madge and Her Friends.
The Children of the Court.
Maid Marjory.

Peggy, and other Tales.
The Four Cats of the Tippertons.
Marion's Two Homes.
Little Folks' Sunday Book.
Two Fourpenny Bits.
Poor Nelly.
Tom Heriot.
Through Peril to Fortune.
Aunt Tabitha's Waifs.
In Mischief Again.

Books by Edward S. Ellis. Illustrated. Cloth, 2s. 6d. each.

Shod with Silence.
The Great Cattle Trail.
The Path in the Ravine.
The Young Ranchers.
The Hunters of the Ozark.
The Camp in the Mountains.
Ned in the Woods. A Tale of Early Days in the West.
Down the Mississippi.
The Last War Trail.
Ned on the River. Tale of Indian River Warfare.

The Phantom of the River.
Footprints in the Forest.
Up the Tapajos.
Ned in the Block House. A Story of Pioneer Life in Kentucky.
The Lost Trail.
Camp-Fire and Wigwam.
Lost in the Wilds.
Lost in Samoa. A Tale of Adventure in the Navigator Islands.
Tad ; or, "Getting, Even" with Him.

The "World in Pictures." Illustrated throughout. *Cheap Edition.* 1s. 6d. each.

A Ramble Round France.
All the Russias.
Chats about Germany.
The Eastern Wonderland (Japan).

Glimpses of South America.
Round Africa.
The Land of Temples (India).
The Isles of the Pacific.
Peeps into China.

The Land of Pyramids (Egypt).

Half-Crown Story Books.

In Quest of Gold; or, Under the Whanga Falls.
On Board the *Esmeralda ;* or, Martin Leigh's Log.
The Cost of a Mistake.
For Queen and King.
Esther West.

Three Homes.
Working to Win.
Perils Afloat and Brigands Ashore.
Pictures of School Life and Boyhood.
At the South Pole.
Ships, Sailors, and the Sea.

Books for the Little Ones. Fully Illustrated.

Rhymes for the Young Folk. By William Allingham. Beautifully Illustrated. 1s. 6d.
Cassell's Robinson Crusoe. With 100 Illustrations. Cloth, 3s. 6d; gilt edges, 5s.

Cassell's Swiss Family Robinson. Illustrated. Cloth, 3s. 6d. ; gilt edges 5s.
The Sunday Scrap Book. With Several Hundred Illustrations. Paper boards, 3s. 6d. ; cloth, gilt edges, 5s.

The Old Fairy Tales. With Original Illustrations. Boards, 1s.; cloth, 1s. 6d.

Albums for Children. 3s. 6d. each.

The Album for Home, School, and Play. Containing Stories by Popular Authors. Illustrated.
My Own Album of Animals. With Full-page Illustrations.

Picture Album of All Sorts. With Full-page Illustrations.
The Chit-Chat Album. Illustrated throughout.

Cassell & Company's Complete Catalogue *will be sent post free on application to*

CASSELL & COMPANY, LIMITED, *Ludgate Hill, London.*

1483260R0

Printed in Great Britain by
Amazon.co.uk, Ltd.,
Marston Gate.